What **Scientists** Think

'There is a powerful and revealing means to find out what scientists think. You ask them.'

From the Foreword, by Marek Kohn

'A pleasure at last to get access to the varied views of scientists.'

Lewis Wolpert, University College London

What are scientists working on today? What do they worry about? What do they think about the working of the brain, climate change, animal experimentation, cancer research, and mental illness? Is science progressing or in retreat? Could this century be humankind's last?

These are just some of the compelling and provocative questions tackled here by twelve of the world's leading scientists and scientific thinkers. In engaging and lucid conversations with Jeremy Stangroom, they clarify many of the most urgent challenges and dilemmas facing science today.

With a Foreword by Marek Kohn, author of *A Reason for Everything: Natural Selection and the British Imagination*.

Jeremy Stangroom is co-editor of *The Philosophers' Magazine*. He is also joint editor of *What Philosophers Think* and *New British Philosophy*.

What **Scientists** Think

Jeremy Stangroom

Routledge
Taylor & Francis Group

LONDON AND NEW YORK

First published 2005
by Routledge
2 Park Square, Milton Park, Abingdon, Oxon OX14 4RN

Simultaneously published in the USA and Canada
by Routledge
270 Madison Ave, New York, NY 10016

Routledge is an imprint of the Taylor & Francis Group

Typeset in DIN and Minion by RefineCatch Limited, Bungay, Suffolk
Printed and bound in Great Britain by TJ International Ltd, Padstow, Cornwall

British Library Cataloguing-in-Publication Data
A catalogue record for this book is available from the British Library

Library of Congress Cataloging-in-Publication Data
What scientists think / Jeremy Stangroom.
        p. cm.
1. Life sciences—Popular works.   2. Science—Popular works.   3. Scientists.
I. Stangroom, Jeremy.
    QH309.W44 2005
    570—dc22   2004025418

ISBN 0-415-33426-8 (hbk)
ISBN 0-415-33427-6 (pbk)

Contents

Marek Kohn

Scientists think in all sorts of ways that keep their thoughts to themselves. They think in depth, in detail and within elaborate frameworks. Often they think in numbers, and sometimes counter to intuition. They are obliged in their intellectual endeavours to be formal, technical and focused. Nevertheless, there is a powerful and revealing means to find out what scientists think. You ask them.

The modesty of this unpretentious technique is its strength. By requiring simplified answers, simple questions encourage a paradoxical kind of precision. They call for responses that are true both factually and as representations of what the responder thinks is worth saying. In other words, they get to the point. A non-expert reader can take away the gist of key ideas and a sketch of the thinker's intellectual map, thereby avoiding errors both of fact and of emphasis. The results are cumulative, too: Jeremy Stangroom's interviews add up to a deceptively compact pocket guide to many of the hottest spots in contemporary science.

These spots are hot not just because of the scientific activity around them, but because of the concerns and values cooking away within them. This collection takes in its stride heredity, human nature, the nature of being human, consciousness, race, mental disorder, experiments on animals, genetic modification of organisms and the extinction of species. It also considers cancer and viruses, reminding us that scientific goals are sometimes fervently desired and universally shared. Taking the broad view, what scientists care about is what everybody cares about.

And what they think is, in many respects, what most people think. Scientists don't usually derive their values from their science; they already have them when they become scientists. They understand issues through their knowledge of science, but their views may not be greatly altered by their knowledge. As a neuropharmacologist, Susan Greenfield is ideally placed to appreciate the effects of psychoactive drugs at a neuronal level. But although her vision of the issue is based on a vision of neural processes, her views on the pursuit of ecstasy – that it should be done in a respectable fashion and in moderation – belong to a moral tradition which has no need of science to reach its conclusions. The familiarity of these views illustrates that much of what scientists think about the ethical implications of their knowledge is accessible to all, and therefore that the barriers to dialogue between scientists and the rest of society may be partly illusory. In ethics scientists are usually amateurs. Whether or not one still considers them to have priestly qualities, outside their specialisms they belong in the pews with the rest of the laity.

Expertise is inevitably exclusive, though. Norman Levitt, a mathematician, points out that to understand the disputes between Newton and Leibniz it is necessary to know calculus. Levitt's standard for the competence required to take an active part in the discussion of scientific issues is high. It implies that outsiders from other scholarly disciplines be not only expert in their own fields but also skilled in those of the scientists they wish to discuss, a criterion that favours basically internal investigations. In the wider public sphere, it suggests a state of affairs in which scientists explain and everybody else listens.

As Jeremy Stangroom observes in his introduction, it is indeed important that people should be paying attention. He finds data on public ignorance of science that is beyond caricature – a third of people still under the impression that the Sun revolves round the Earth, which would seem to justify putting the 'Copernican revolution' in quotation marks. It may be tempting to dismiss this as the shortcomings of classes who have failed, whether through their own fault or otherwise, to be educated. But the managerial and professional classes, membership of which implies a developed capacity for rational thought, are the core market for non-rational therapies such as homeopathy, which is based on the magical belief that 'like cures like' and purveys preparations which are unlikely to contain a single molecule of the supposedly effect- ive agent. A handful of consumers may cite literature reviews in sup- port of their choice, but for most of them logic is not an obstacle. It's not just that people can't be bothered to inform themselves about basic science, but that they feel free to choose the bits they like and reject the ones they don't. The selective attention given in some quarters to minority views of evolution is a case in point.

Scientists are not above inconsistency themselves, though. There may be one or two 'yes, but' moments for readers of these interviews; and it may be easy to see the joins between the scientific observations and the social comments. These can be seen as opportunities rather than flaws, however: openings for non-scientists to engage actively rather than to absorb passively. And while Norman Levitt may be right about professional standards, the cosmologist Martin Rees is surely also justified in his confidence that most key scientific facts and concepts can be conveyed effectively in simple language. Professional knowledge may be the preserve of a very few, but working knowledge of science is within nearly everybody's grasp. In seeding such a knowledge, this informative and stimulating collection might help relieve that patrician burden, the 'public understanding of science', with a civic alternative: public engagement with science.

The Scientists

Colin Blakemore is the Chief Executive of the Medical Research Council, a post he took up in 2003. He is also Waynflete Professor of Physiology at Oxford University, a position he has held since 1979. He has been President of the British Neuroscience Association, the Physiological Society and the new Biosciences Federation. He has also been President and Chairman of the British Association for the Advancement of Science. His radio and television work includes *The Mind Machine*, a thirteen-part series on brain and mind, broadcast on BBC2 in 1988. His book *Mechanics of the Mind* won the Phi Beta Kappa Award in Science.

Dorothy Crawford is the Robert Irvine Professor of Medical Microbiology at the University of Edinburgh. Her primary research interest is the Epstein–Barr virus (EBV), a ubiquitous human herpes virus which persists in the body for life after an initial infection. She is the author of *The Invisible Enemy: A Natural History of Viruses*.

Susan Greenfield is Fullerian Professor of Physiology and Director of the Royal Institution of Great Britain, and Professor of Synaptic Pharmacology at Oxford University. She was an undergraduate and graduate student at Oxford University, and she did postdoctoral research at College de France, Paris, and at the New York University

Medical Center. She presented the BBC television series *Brain Story* in 2002, and has written several books for the general reader, including *The Private Life of the Brain* and *Tomorrow's People*. She was awarded the CBE in the Millennium New Year's Honours List and a Life Peerage in 2001.

Steve Jones is Professor of Genetics at University College London. He gave the 1991 Reith Lectures on 'The Language of the Genes'; wrote and presented a long-running BBC Radio 3 series on science and the arts, *Blue Skies*, and a six-part TV series on human genetics, *In the Blood*, which was broadcast in 1996. His books include *The Language of Genes*, winner of the prestigious Rhône-Poulenc Science Book Prize, and *Almost Like a Whale: The Origin of Species Updated*. He is the President of the Galton Institute.

Norman Levitt is Professor of mathematics at Rutgers University. He is the author of *Higher Superstition: The Academic Left and Its Quarrels with Science* (with Paul R. Gross) and *Prometheus Bedeviled: Science and the Contradictions of Contemporary Culture*; and the co-editor of *The Flight from Science and Reason*.

Kenan Malik is an author and broadcaster. He studied neurobiology at the University of Sussex, and History and Philosophy of Science at Imperial College London. In between, he was a research psychologist at the Centre for Research into Perception and Cognition (CRPC) at the University of Sussex, working on problems of the mental representation of spatial relations. His books are: *The Meaning of Race* and *Man, Beast and Zombie: What Science Can and Cannot Tell Us About Human Nature*. He is one of the presenters of BBC Radio 4's *Analysis* programme. He is a fellow of the Royal Society of Arts.

Robin Murray is Professor of Psychiatry and Head of the Division of Psychological Medicine at the Institute of Psychiatry. He received his medical degree from the University of Glasgow, Scotland, and took his Boards in Internal Medicine there in 1972. He then trained at the Maudsley Hospital in London and has remained there ever since, apart from one year as MRC Fellow at NIMH. Between 1994 and 1996, Professor Murray held the Presidency of the Association of European Psychiatrists.

Steven Pinker is the Johnstone Family Professor of Psychology at Harvard University, a post he took up in 2003. Prior to this he spent twenty-one years teaching at MIT. He is the author of a number of best-selling books, including *The Language Instinct, How the Mind Works* and *The Blank Slate*. His research on cognition and language won the Troland Award from the National Academy of Sciences and two prizes from the American Psychological Association.

Martin Rees is Professor of Cosmology and Astrophysics and Master of Trinity College at the University of Cambridge. He also holds the honorary title of Astronomer Royal. He is a foreign associate of the National Academy of Sciences, the American Philosophical Society, and is an honorary member of the Russian Academy of Sciences. He is the recipient of many awards, including the Gold Medal of the Royal Astronomical Society, and has written five books for a general audience, including *Our Final Century* and *Just Six Numbers*.

Mike Stratton is the Head of the Cancer Genome Project at the Wellcome Trust Sanger Centre in Cambridgeshire. Professor Stratton established the Cancer Genome Project, with Dr Richard Wooster, at the end of 1999, with the aim of using data from the Human Genome Project in order to pinpoint the genetic abnormalities associated with all cancers. Earlier in his career, he headed up the team at the Institute of Cancer Research which discovered the BRCA2 breast cancer gene. He retains his links with the Institute, chairing the Section of Cancer Genetics.

Kevin Warwick is Professor of Cybernetics at the University of Reading. He began his career at British Telecom at the age of 16, and spent six years there, before embarking on a degree at Aston University. He followed this up with a doctorate in computer engineering and robotics at Imperial College London, where he went on to hold a research post. Subsequently, he held positions at Oxford, Newcastle and Warwick universities, before taking up the Chair at Reading at the age of 32. He is the author of a number of popular science books, including *In the Mind of the Machine* and *I, Cyborg*.

Edward O. Wilson is Professor of Science and Curator of Entomology at the Museum of Comparative Zoology at Harvard University. He is the recipient of numerous scientific awards, including the National Medal of Science and the Crafoord Prize of the Royal Swedish Academy of Sciences. Two of his books, *On Human Nature* and *The Ants*, have won Pulitzer Prizes. He is a member of the National Academy of Sciences, and is a Fellow at the American Academy of Arts and Sciences and the American Philosophical Society. *Time* magazine named him one of America's twenty-five Most Influential People in 1996, and in 2002 presented him with a Lifetime Achievement Award.

Preface

I would like to offer thanks to the following people for their help in putting this book together: Tony Bruce, for a good idea, and for editorial help; Ophelia Benson, my colleague at ButterfliesandWheels.com, for reading and commenting on the first draft of the book; John Matthews for library assistance; Matthew Broome for answering my questions about psychiatry; Paul Allen for tracking down a particularly elusive reference in the psychiatry literature; the kind woman who drove me three miles to the Sanger Centre at Hinxton, after finding me wandering around lost in the Cambridgeshire countryside; and last, but not least, the twelve interviewees for giving up their time for this project.

Jeremy Stangroom
London, August 2004

Introduction

Jeremy Stangroom

In March 2002, the Royal Society, the independent UK academy of sciences, held a national forum to discuss whether scientists were trusted by the general public. In his opening address to the forum, Sir Paul Nurse, Chairman of the Royal Society's Science in Society Committee, argued that public confidence in science had been shaken by the controversies surrounding issues like global warming, biological weapons, GM food, BSE/CJD, animal experimentation and nuclear power. This view is borne out by opinion poll data. For example, a recent MORI poll found that 48 per cent of people, compared with 36 per cent in 1989, agreed that scientists don't know what they're talking about when it comes to the environment; and the same poll found that scientists ranked below doctors, teachers, judges and clergy in terms of how far they were trusted to tell the truth.

In conjunction with this lack of confidence in science, the public displays a fairly alarming lack of knowledge about basic scientific facts. For example, survey data in the UK and the United States show amongst other things: that some two-thirds of people are unable to name CO_2 as the main greenhouse gas; that up to one-third of people think that the Sun travels around the Earth; that more people than not think that antibiotics kill viruses as well as bacteria; and that one in five people are unable to say anything at all about the meaning of the phrase 'human genetic information'.

The response to this situation on the part of government, media commentators and representatives of the scientific community has been to argue that there is a necessity for scientists to engage the public in a dialogue about their work. For example, Professor David King, the UK government's Chief Science Adviser, has argued that honesty and openness is the key to recovering public confidence in science and in science policy making; and Charles Secrett, one time director of Friends of the Earth, has claimed that people would have more trust in scientists if there were greater freedom of information.

The rationale for this book of interviews is predicated on the idea that there is something right about this notion that scientists should seek to communicate their views to the wider public. In particular, there are two reasons why an increased public understanding of science is desirable. The first has to do with the fact that whether science flourishes or not is partly determined by the social and political context in which it takes place. Therefore, there is a necessity for scientists to engage with the public in order to gain broad support for what they are doing. The debate over GM foods in the UK is instructive in this respect. To date, the advocates of GM foods have been unable to make any headway persuading the British public of their potential benefits. As a result, in the face of consumer resistance and direct action campaigns organised by environmental pressure groups such as Greenpeace, biotechnology companies Monsanto, Syngenta and DuPont have all cut back their British-based GMO (Genetically Modified Organisms) research.

This is not to argue that an increased public understanding of science will necessarily result in support for any particular scientific endeavour. However, there *is* evidence to suggest that the public will support a scientific project which has clear benefits; and that where there is a lack of support for any particular project, it is frequently because people do not understand where the benefits lie. Again the debate about GM technology is instructive. At least some of the public resistance to GM foods is rooted in the belief that they are bad for human health, yet there is no evidence to suggest that this is the case. It is also likely that the campaign of disinformation perpetrated by the various environmental pressure groups has blinded people to the potential benefits of GM foods, not least the possibility that if crop yields can be increased significantly, then more people will get fed.

The second reason for thinking that an increased public under-standing of science is desirable has to do with a commitment to the value of truth. If there is something *intrinsically* valuable in the accumulation and dissemination of scientific knowledge, then a scien-tifically literate population is a good thing. Of course, it is likely that increased scientific literacy will also bring purely extrinsic benefits; there are good reasons, for example, for thinking that scientific literacy will be correlated with an ability for a certain kind of critical thinking. Nevertheless, it remains possible to argue that scientific literacy is desirable in and of itself; that part of what is special about the human species is that it is able to find out things about the world, and that science, via its very particular method, is the best way of going about this task; and that if people, content to sleepwalk their way through their lives in the comfort of myth and superstition, forgo this aspect of their humanity, then they are turning their back on a large part of what has been bequeathed to them by their evolutionary heritage.

The rationale for this book then is two-fold. First, it is hoped that it will show why the work which scientists do is important and why they deserve to be supported in their endeavours. And second, it is part of an on-going process of the dissemination of scientific knowledge, something which is desirable in and of itself.

The Interviews

The book comprises twelve interviews with leading scientific thinkers. Not all, strictly speaking, are scientists – Norman Levitt, for example, is a mathematician, and Kevin Warwick works at the intersection of science and technology; however, they all have interesting things to say about science, scientific method and science in society. Their inclusion in the book has not been determined simply by their standing in the academic community; all the interviewees, of course, have considerable reputations, but, in every case, at least arguably, it would have been possible to have chosen another person with an equal reputation. Rather, what every person here offers is expertise and interest in areas of scientific endeavour which have particular social, political and ethical dimensions.

This reflects the intention that this book should explore some of the issues which emerge as a result of the fact that science, in the form that it takes and in its effects, is thoroughly enmeshed in the social world. Specifically, we're talking here about issues like: animal experimentation; bioterrorism; GM technology; the scope of evolutionary explanations; machine intelligence; the accountability of science to the wider public; anti-science, in its postmodernist and religious guises; medical advances; the erosion of biodiversity; and so on.

One consequence of the concern with these kinds of issues is that the focus of the book is slanted towards the life sciences. Indeed, there are important areas of enquiry which receive no extended discussion in these interviews at all; the field of theoretical physics is perhaps the most significant example. Needless to say, this is not to suggest that some parts of science are either more significant, or more interesting, than other parts. It is rather a function of the fact that the significance, in terms of how people live their lives, of something like mathematical physics is by an order of magnitude more nuanced than in the case of, say, evolutionary biology, and it is, therefore, much less likely to give rise to the kinds of social, political or ethical concerns which provide the focus of a large part of this book. It is also a function, it must be said, of the nature of the interview format: the more abstract the science, the less suitable it is for treatment in the context of a dialogue, especially when one of the participants is not a professional scientist.

How to Read this Book

The interviews have been arranged so that they progress thematically. The reader, therefore, will get most out of the book if they follow the instructions given to the White Rabbit in *Alice's Adventures in Wonderland*: begin at the beginning, go on until the end: then stop. It *is* possible to treat each interview as a separate entity, and to read just those which are of specific interest; however, doing so will make it harder to discern how various common themes and threads connect up over the course of the volume.

The style of the interviews is deliberately conversational. To facilitate the flow of the discussion, the interviewees were not asked to explain every time they used a piece of technical language. Nevertheless, every effort has been made to make it possible for people without prior knowledge of the subject area to follow the discussion, so whilst readers might miss the occasional allusion, the broad thrust of what is being said should always be comprehensible.

A Final Word

At the beginning of this Introduction, it was stated that there is *something* right about the idea that scientists should seek to communicate their views to a wider public. There is also something wrong with this idea; or, at least, something wrong with one of the unspoken assumptions which underpins it. To understand what, it is worth considering the following story. After the UK General Election in 2001, the media were full of commentary bemoaning the low turnout, and putting the blame squarely at the feet of the politicians. People were put off politics, it was said, because it was all about spin, or it was boring, or it was just point scoring, and all politicians were just in it for themselves, anyway. This is what Jackie Ballard, ex-MP for Taunton, had to say about all this:

> Blame the politicians? It's time somebody blamed the voters. Those of you who can't be bothered to walk down the road once every couple of years and put a cross on a ballot paper. . . . Many people around the world would literally die to get the right to vote in a free election. There's a hell of a lot of things wrong in this country and a lot which needs changing . . . but if you just sit at home moaning about it you will get the Government and the politicians you deserve.

The point, of course, is that science and scientists find themselves in a position similar to that of politicians. If people are turned off science, then, we are told, there must be something wrong with science, or with the way in which scientists engage with the wider public. But this just isn't clearly the case. Many scientists *already* go out of their way to explain their work to a general audience. Consider, for example, that popular science has been one of the publishing success stories of the last twenty years; or that there are myriad public events which deal with scientific themes; or that it is possible to pick up at least three different science periodicals from the news-stands; or that the Internet features numerous web sites devoted to the task of making science accessible; or that terrestrial television frequently features programmes on scientific topics.

Not one scientist declined to be interviewed for this book. Scientific information is readily available to the general public. In the end, people have to take some responsibility for keeping themselves informed. If they do not, then probably they'll get the science that they *don't* deserve, because science tends to deliver major improvements in the quality of the lives of human beings. However, if things go badly in the twentieth-first century – if, for example, as Martin Rees thinks possible, we are struck by some kind of technological disaster – then it isn't clear that people, who, for no good reason, have not bothered to find out that the Earth goes around the Sun, are in any position to complain too loudly about their fate.

1 Darwinism and Genes

In conversation with Steve Jones

In his book *The Blind Watchmaker*, Richard Dawkins argues that prior to 1859 it was not really possible to be an intellectually fulfilled atheist. The difficulty for atheism was that the complexity of the living world, in the absence of an alternative explanation, suggested the hand of a designer at work. However, this all changed in 1859 with the publication of Charles Darwin's *The Origin of Species*, which brought to the world's attention perhaps the most powerful single idea in the history of science: namely, the idea of the evolution of species by the mechanism of natural selection.

Broadly speaking, this idea has it that natural selection works on the variations in the inherited traits of organisms, by ensuring that those traits which are most likely to promote reproductive success – for example, by helping an organism stay alive long enough to breed – are those which are most likely to be passed on to offspring, and which will, therefore, become predominant. However, in 1859, a problem for Darwin's theory was that the mechanisms of inheritance were not yet understood.

'Darwin got inheritance fundamentally wrong,' says Steve Jones, Professor of Genetics at University College London (UCL). 'In fact, it's actually rather hard to know precisely what he thought about it. Nineteenth-century biologists weren't really that interested in theories of inheritance. They were more interested in development; how, for example, an elephant's egg, which is a small formless piece of proto-plasm, develops into an elephant.

'But Darwin did have this idea that there were gemmules, bits of information, which floated in the blood, and which were mixed at conception and then passed on. But after the publication of the first edition of the *Origin*, a Scottish engineer, Fleeming Jenkin, wrote to him, and pointed out, roughly speaking, that if you mix red and blue, you get purple. You can never get either red or blue back again. So if it were advantageous to be red, the advantage would be lost as soon as it was mixed; any advantageous character would be diluted out.

'Darwin was that rare thing, an honest scientist, and he saw at once that this was a fatal flaw in his argument. So you find that the later editions of the *Origin* are less sure, less mature. But, of course, it turned out that the flaw wasn't a flaw at all. Just a couple of years after the *Origin*, Mendel showed that inheritance was based on particles, not fluids, and that inheritance is detached from the physical structure of the people who pass on the genes. And this rescued Darwinism, because you can get the red gene back, even from a purple individual.'

We now know that genes are the units of biological inheritance. People often talk as if genes code for particular characteristics in a straightforward way. So, for example, the media have made much of the possibility of a 'gay gene'; and people will talk about genes for particular illnesses, for instance a gene for schizophrenia. But aren't things actually a lot more complicated than that?

'Yes, they are,' Jones replies. 'In some ways, this is to go back to the nineteenth-century question about development; it is about the relationship between genotype and phenotype. Every geneticist knows that there is not a straightforward mapping between genotype and phenotype; we know this so well that we don't talk about it. But the

general public don't realise it, and they use the most dangerous word in
genetics, which is the three letter word, 'for': a gene for this and a gene
for that.

'There *is* something right about this way of talking. If, for example,
a person has one particular DNA chain, then they will grow up as an
achondroplastic dwarf. In shorthand, that DNA chain is a gene for
dwarfism. But this is unusual. For most of the attributes which we find
interesting – for example, things like personality and appearance –
there are very few single genes where there will be a direct relationship
between a genetic change and a phenotypic change.'

If that's the case, how is it possible to specify an individual gene?
What actually gives a particular DNA chain its reality as a gene?

'That is not an easy question; it's a technical issue, and there isn't
necessarily a right answer,' says Jones. 'Roughly speaking, a gene can be
defined as a unit of information. That's how Mendel saw it. Biochemi-
cally, it's a segment of DNA, which produces an RNA message which is
read off it. But now that we know more about the genome, the old
central dogma, that genes make RNA makes protein, is by no means
straightforward. For example, there are genes within genes, some genes
overlap with each other, and there are great acres of DNA, called "junk
DNA", which doesn't seem to do much at all. So "gene" is a convenient
shorthand for what may turn out to be a rather complicated
phenomenon.'

Are those philosophers who claim that there is a lot of confusion in
the use of the word 'gene' then right in their criticisms?

'I don't think biochemists are going to be the least bit interested in
what philosophers think about genes,' Jones replies. 'As I've said in the
past, philosophy is to science as pornography is to sex: it's cheaper,
easier and some people prefer it. But there is this rather philosophical
point in biology that the more you find out, the more complicated

things get. We're very much in the position of finding out stuff, and then finding the exceptions to what we think we know. So at the moment what a gene is isn't exactly clear.'

The language of 'genes for' particular characteristics causes quite a lot of friction. Both Richard Dawkins and Ed Wilson, for example, have felt the need to defend its use. Does Jones think that it is a sensible way of talking or does it involve the potential for too much misunderstanding?

'I think it is a useful shorthand, and so long as you have your fingers crossed when you use it, then it's probably OK,' he answers. 'There are certainly genes for lung cancer, for example. If everybody smoked, then lung cancer would be a genetic disease; a person's ability to cope with the poisons in tobacco is very much a matter of the genes which they've inherited. But it would be foolish to say that the genes for lung cancer are independent of people's smoking habits. They interact. So the word "for" has a nimbus of other meanings around it. Matt Ridley, in his book *Nature via Nurture*, makes the point that geneticists don't actually talk about the nature/nurture controversy, because they know it's not a controversy; there is both nature and nurture, and it is very difficult, if not impossible, to untangle the two.'

In his book *Almost Like a Whale*, Jones argues that 'every part of Darwin's thesis is open to test. The clues – from fossils, genes or geography – differ in each case, but from all of them comes the conclusion that the whole of life is kin.' How, I ask, does our understanding of genes contribute to this conclusion? Is the key point that we can identify genes, or at least chains of nucleic acids, which different species or organisms share in common?

'Well, that's one of the key things, but there are others,' he replies. 'Modern molecular genetics is really just comparative anatomy plus a huge amount of money. Darwin's arguments were underpinned by his undoubted skills as an anatomist. He looked, for instance, at many kinds of barnacles, and he found that despite a huge variation in their appearance, they are built fundamentally the same way. With much less skill you can now do this with any group of creatures. For example, if you're looking at the genes which increase life expectancy in the drosophila, it is possible to go to a database to look for the same genes in humans, and what you find is that in humans they function to repair DNA damage. So you know that life expectancy and DNA damage are related. This tells you that the plans for building things as different as a fruit fly and a human being are actually quite similar. So what genetics has done is to introduce improved weapons to the Darwinian argument.'

A resolute anti-Darwinian, of course, is unlikely to be convinced by this line of argument. If they were a creationist, for example, then likely they'd say that it would be expected that God would use similar mechanisms to build different organisms.

'That's a different kind of point,' says Jones, when I put it to him. 'I'm like Richard Dawkins in that I won't debate with creationists, because they're not engaging in rational argument. Certainly it is possible to say, "Well, God would have done that", but that's the equivalent of Philip Gosse's argument that fossils might have been put into rocks by God in order to test the faithful. Well, maybe that's right, but until there is some evidence to support the view, I'm not going to believe it.'

A different kind of challenge to Darwin's idea of natural selection comes from those people who claim that the fossil record suggests that evolution proceeds in fits and starts rather than by means of slow, gradual change.

'This is a real argument, and happily it's one which takes place between biologists,' Jones says. 'Steve Gould wrote a paper called "Is a new and more general theory of evolution emerging", in which he pointed out that even the best fossil records show apparent leaps from one form of organism to the next. Darwin had made a real song and dance about the fact that nature does not make leaps. He did this in order to avoid the creationist model which allows that a species might turn instantly into another species. Darwin's claim was that the incompleteness of the fossil record explained the kind of thing which Gould identified. But Gould insisted that it was a real phenomenon; that once you get a particular form, it is very hard to get from that form to another one. You had to go through a revolution, which would involve all kinds of disruption, gene interactions, bottlenecks, and so on. This is an interesting argument. Gould certainly made the point that people's blind assumption that the best fossil records were always gradual was just wrong. It is clear that they do show discontinuities.

'However, there's a two-part difficulty with the argument. The first is that what is a jerk to a palaeontologist like Gould is a creep to a biologist. So there's a mind-set difference. And second, even the most traditional of Darwinists don't deny the fact that some things stay the same for a long time, especially where the environment stays the same.'

Despite overwhelming evidence for the truth of Darwinism, opinion surveys in the United States consistently show that less than 50 per cent of people think that human beings are evolved animals. What does Jones think of the situation in the States? Is it a cause for serious concern?

'There is a kind of new ignorance out there which is depressing,' he replies. 'There is a deep intellectual dishonesty about creationism which I really despise. There is dishonesty in that they lie to themselves. But what's more important is that they're lying to a new generation of children. They're using a false claim of intellectual honesty to spread intellectual dishonesty.

'You don't start every geography lesson with a proof of the fact that the Earth is round; we've gone beyond that. And we should have gone beyond having to prove Darwinism every time we talk about evolution. But in the US, a lot of biologists in universities have to spend a large part of their time putting the kids right about the fundamental facts of biology, things which they should have been taught at school, but weren't.

'Although philosophy is largely useless, I think there is some interesting philosophy of science. I've always liked Karl Popper's view that science is about disproving things. If we find Neanderthal man wearing a Rolex watch, then I'm going to give up the theory of evolution. But so far that evidence has not appeared. Or take the fact that the Sun is at the centre of the Solar System. If astronomers were suddenly to find that the Sun went around the Earth, they might moan and groan, but they would admit that they were wrong. They've done this before, in 1905, with Einstein's theory of relativity; they threw everything out of the window, and effectively started again. But creationists don't think like this. They think that unless it is possible to prove Darwinism beyond all doubt, then it isn't true. But ironically, of course, what they've committed to themselves is both entirely beyond proof and also unfalsifiable.

'It's also interesting that if you look at the literature on evolution, it is full of disagreement. Indeed, there is a lot of bitter disagreement: creeps and jerks, as we discussed, for example. Biologists spend their time quarrelling. If you go to a church, you just don't see the same kind of disagreement. They all think God is marvellous, and that Jesus was a great bloke. So, if you're very naïve, it sounds as if the creationists have a stronger case than we do, when, of course, the opposite is the case.'

It isn't only creationists who are made nervous by talk of Darwinism and genes. Many people working in the humanities and social sciences, for example, are suspicious of theories which appeal to innate factors in order to explain particular behaviours or particular characteristics of the social world. This becomes especially controversial when it involves the comparison of different groups of people with each other. The classic example of this is the in-fighting which has surrounded those studies which have compared the IQ levels of different racial groups, and which have found that there are systematic differences. What does Jones make of this kind of work? Does it have scientific merit?

'This has always been a messy issue, because obviously there are political implications,' he answers. 'This sort of research has been used, post-Galton, for various things which we don't even want to think about. But nevertheless, there is a scientific issue here which is worth considering. It isn't out of the question that one population group is smarter than another for genetic reasons.

'There has been a lot of debate in the United States recently about the use of racial categories in medicine. In the literature, there has been much discussion about whether black people respond better than other groups to particular drugs or treatments. A recent paper looked at all the studies. The first thing that was found was that skin colour is a proxy for poverty. This is so obvious that people tend to forget it. But if you take it away, whilst the difference between racial groups narrows, it is still there. In particular, there's a heart medicine which apparently works better for blacks than it does for other people. And this sparked a controversy. When it was first suggested, three or four years ago, a whole load of angry lefties wrote to say that people shouldn't say this kind of thing. But this is silly; if there are differences, then we need to know.

'But what was really interesting was that when a group here looked at differences around the world, they found that whilst there were indeed differences in how population groups responded to particular drugs, these differences did not coincide with traditional racial classifications. For example, West Africans don't respond in the same way as East

Africans. So the traditional racial groups, which appear to the naked eye to be distinct, perhaps rather surprisingly are not.'

'Even so,' Jones continues, 'it is still possible that there are genetic differences between groups which have an effect on IQ scores. But actually the data does not support this conclusion. Let's assume that the heritability of IQ is 0.6; that is, that the proportion of total variation in IQ due to genetic variation is 0.6. Let's accept, though I wouldn't, that there is a difference of 15 points between the IQs of black and white Americans. It might seem to follow that given that most of the variation in IQ is due to inherited factors that this difference between black and white scores must be due to genetic factors. But this is a fundamental misunderstanding of the meaning of the word heritability.

'The way I often illustrate this is to talk about blood pressure. If you measure the blood pressure of middle-aged black men in the US, you find that, on average, they have a higher blood pressure than white men. This causes quite a lot of mortality. If you look at the heritability of blood pressure it's around 0.4. So doctors then say, "Oh dear, it's the lifestyles of black people, they smoke too much, they don't do enough exercise, their environment must be pushing up their blood pressure. We should do something about this." But, on the basis of exactly the same data, educationalists say, "Oh, it must be genetic, there is no point in doing anything about it." But, of course, they're both wrong. We know that there is some genetic predisposition among some Africans towards high blood pressure; they have high levels of salt in their blood, which pushes up their blood pressure. But there are also environmental factors – too many cheeseburgers, for example – which increase their blood pressure.

'So the data simply doesn't support the case that any difference in the IQ levels of different racial groups is a function of genetic differences. Geneticists have made this point over and over again, but still some people refuse to accept it.'

Perhaps the aspect of the science of genetics which worries people most in the UK has to do with genetically modified crops. However, this worry does not seem to be shared by most scientists. Ed Wilson, for example, argues that whilst there are legitimate concerns about GMOs, the potential benefits outweigh the risks, so we should continue to use the technology. What's Jones's view about this?

'Actually, I think that the whole GM debate in the UK has been lamentable,' he replies. 'First of all, there was the blithe assumption on the part of scientists that they could go ahead without consulting the public. This, of course, just shouldn't be true; I don't think any serious scientist would ever claim that it ought to be true.

'The Greens for their part have just been totally dishonest about the whole thing. A little while ago, I was booed by 3,000 people at Central Hall in Westminster, after I asked which political party had been the greenest in Europe; there's no doubt about the answer, it was the Nazis. They were completely green. They were interested in the purity of blood. They didn't like smoking. They wanted forests. They were all for exercise. They didn't like cars. A lot of this anti-GM food stuff has a strange atavistic purity of blood thing going on. I don't want to overstate it, but it is definitely there.

'The Greens have been absolutely cynical in manipulating public concern about GM foods. It's the dishonesty of their approach which annoys me. If they don't like industrial agriculture or the common food policy, then I'm in agreement with them. But to muddy the water with completely spurious arguments about GM food and human health, for example, is a disgrace.'

So does Jones then share Ed Wilson's view that it is the potential for increased crop yields which makes it worth continuing with GM technology?

'Well, potential is the key word here,' he answers. 'In general terms, genetic modification has not paid off as it was thought it might. There have been increased yields, but they haven't been sustained. The one major exception is cotton, where the technology has been successful. In China, for example, GM cotton is almost universal, so the simplistic view is that it is clearly more cost-effective than traditional cotton. But the argument which the Greens make is that the cost–benefit calculations do not include secondary costs; for example, the chance that peasant farms will be lost.

'However, in the US, this has happened already with the development of hybrid corn, which had a dramatic effect. It meant that farmers could no longer take the seeds from their own crops to plant for the next generation. They had to go back to the seed supply companies. These companies were ruthless; they forced small farmers out of business, and now there is this seamless agribusiness, with small farmers having largely disappeared.'

'But, having said this, the US agricultural business is the most efficient in the world,' Jones says. 'Basically, it feeds the world. So do you want to feed everybody or do you want to have decorative peasants on farms? My view is that you need to feed the world. It just isn't the case that small farming is the most effective way to grow food, and it's intellectually dishonest to say it is.'

Jones's suggestion that the immediate benefits of GM crops have been oversold matches a criticism which has been made of some of the hype surrounding the Human Genome Project. With the completion of the mapping of the human genome, the media made much of the fact that it would lead to new medical techniques. However, it is probably fair to say that within the scientific community people are pretty pessimistic about the time scales involved. The psychiatrist Robin Murray, for example, points out that until very recently the search for genes associated with schizophrenia had always ended in failure. So when might we expect our knowledge of genetics and the human genome to bring us major benefits in terms of health and life expectancy?

'Well, prediction is always difficult, but it is clear that the initial hopes were overstated,' Jones answers. 'Some of this was hype, but some of it was simple optimism. The first schizophrenia gene was found here at UCL in 1991. The trouble was that nobody else was able to find it. However, because of genetics, we now know that schizophrenia seems to be more than one disease. So, in this sense, genetics *has* done quite a lot. Heart disease is another example. Many coronary heart diseases present with the same symptoms; but if you do a genetic test, you can determine the precise cause of the problem, and you can treat it with the appropriate drug.

'So genetics is already an important part of medicine. At the moment, it is more important in diagnosis than it is in treatment. But historically, diagnosis has been the first step on the way to new treatments. So whilst people are more sober than they were about the impact of genetics, it would be wrong to think that it is unlikely to lead to dramatic changes. Pretty much everybody nowadays dies of a genetic disease. If you were to abolish all deaths in the UK before the age of 50, mean life expectancy would only increase by twelve months. In other words, we've won. The enemy without, infectious diseases, accidents, and so on, certainly in the Western world, has been defeated. It may only be temporary, but for now the battle has been won. The enemy within is what kills us. Heart disease, cancer, diabetes, and so on, nearly all have a genetic component. So it would be foolish not to study the genetic basis of these things, because we're not going to conquer them if we don't understand them.'

There is a set of issues here about the kind of knowledge which we might start to have about our own personal destinies in terms of health. For example, it is possible to take a test for Huntington's, an incurable and fatal genetic disease, but not everybody at risk elects to take the test. Presumably this kind of situation is something which we all might face, in one form or another, in the future?

'Yes, but it is already the case that if you walk into a doctor's surgery, you can find out within five minutes how much longer you're likely to live,' Jones says. 'Six factors – age, weight, sex, blood pressure, smoker or non-smoker, and income level – contain about 75 per cent of all the data on life expectancy. At the moment, genetics is irrelevant for 99 per cent of the population. Of course, there will be some people who will find out, after taking a test, that they have, for example, Huntington's. But, at the present time, these people are a very small part of the population. Of course, this part will get bigger, but when you consider that people are willing to smoke, despite the disastrous effect which this has on their life expectancy, they are not likely to be too worried by a DNA test which might tell them that they have a predisposition towards, for example, heart disease.'

Will insurance companies be happy with this situation, though? Presumably, it will make things rather difficult as far as something like health insurance is concerned?

'Insurance companies are in the business of making money,' Jones replies. 'They're interested in knowing more about the odds than their clients know. Life insurance people aren't going to care, because life is a bimodal state, you're either alive or you're dead. With health insurance, it *is* more difficult. If you've got Huntington's, for example, you're going to stay alive for twenty years, and cost a great deal of money. So it's less clear what's going to happen here. It looks like the insurance companies might just take a deep breath and deal with it. The problem is that they know that people who take a test will buy insurance if they're at risk, but they won't if they're not. And this messes up the health insurance industry. So it looks like they might just say that they want to forget about genetic tests. The rule at the moment is that if you've taken a test, you've got to tell them. But this isn't likely to be sustainable into the future, as genetic tests become a standard part of diagnosis. My response to all this is that people should get real. We've all got genetic damage. Is it going to help us to know? Probably not.'

In *Almost Like A Whale*, Jones suggests that it is possible that the human species will disappear. Presumably this is because of the time scales which evolution operates across, and because the history of nearly all species is that they disappear in the end.

'Those things are true, but *Homo sapiens* is the species which, in a sense, has stepped outside evolution, so standard evolutionary predictions don't actually tell us very much,' he says. 'The one thing which is overwhelmingly true about *Homo sapiens* is that in evolutionary terms there are far too many of us. If you draw a line for all mammals showing the relationship between body size and abundance, every mammal will sit on the line, except for *Homo sapiens*, which is 10,000 times more common than would be expected. The population of humans in the world should be equivalent to the London Borough of Camden, though one hopes not the same people! So we're a fantastically abundant species. It is well known that abundant species are liable to epidemics. And we're not only abundant, we're also mobile. So we're definitely at risk from major diseases. But, as I often say, I'm much more scared of Einstein, than I am of Darwin. In other words, I think our political stupidity is more likely to put an end to us than anything biological.'

So is it a possibility then that *Homo sapiens* could buck the evolutionary trend and still be around in a million years' time?

'It is not entirely out of the question,' Jones replies. 'A line I sometimes take is to say that natural selection depends on inherited differences. In the case of humans, natural selection has stopped. Pretty much everybody stays alive long enough to pass on their genes, which means that the opportunity for selection has gone. This may not last, but at the moment it is certainly the case. So *Homo sapiens* could be around for a while yet.'

2   Evolutionary Psychology and the Blank Slate

In conversation with Steven Pinker

If you have studied sociology, the chances are that you have been taught that biology plays little or no part in determining human behaviour. For instance, if there is conflict in society, then it is not to be explained in terms of categories which refer to an innate predisposition towards aggression. Rather, it is to be seen as, for example, a function of a disruption in society's normative system, or the result of a fundamental conflict of interest between opposing social groups.

This kind of marginalisation of biology is predicated on a particular view of the human mind. The Standard Social Science Model (SSSM), as it has been called by Leda Cosmides and John Tooby, holds that the mind has no specific content; it is simply a general purpose machine suited for tasks such as learning and reasoning. Therefore, whatever is found in human minds has come in from the outside – from the environment and the social world. Consequently, if, for example, humans tend to manifest a preference for their in-group, it is because they have learnt that the familiar is good; it is not, for instance, because they have a built-in tendency to favour what they know over what they don't.

However, this model of the mind is falling increasingly into disrepute. The seeds of its current difficulties were sown in the late 1970s with the emergence of sociobiology, which seeks to explain social behaviour in terms of its contribution to the Darwinian fitness of the individuals manifesting the behaviour. However, whilst sociobiology has ushered thoughts about human nature to centre stage, it is evolutionary psychology which poses the biggest challenge to the SSSM. In particular, advocates of evolutionary psychology, such as Steven Pinker, Johnstone Family Professor in the Department of Psychology at Harvard University, argue that the brain is not a general purpose machine, as is supposed by the SSSM, but rather that it comprises a number of functionally distinct parts, which have evolved as solutions to particular problems in our evolutionary past.

What evidence is there, I ask Pinker, for this proposition that the mind is made up of distinct parts?

'The evidence comes from a variety of sources,' he replies. 'For example, the faculty of intuitive psychology is *selectively* impaired in autistic people. A person with autism will fail a false belief test; that is, they won't make correct predictions about a person who has a belief about the world that differs from their own. However, they pass a false *photograph* test; that is, they will know that a photograph can have a depiction of reality that is different from their current knowledge. The formal similarity of these tasks suggests that there is a selective deficit in the ability to conceptualise the content of another person's mind.

'There are other examples of selective impairments. For example, in aphasia, language can be lost while many other aspects of intelligence are spared, including, in some case studies, the ability to solve problems in intuitive psychology such as the false belief task. And there is a genetic disorder called Williams' syndrome which is associated with an inability to conceptualise three-dimensional spatial relationships, but which has far less of an effect on performance in other areas, including language, intuitive psychology and social skills.'

How do people who are committed to the idea that the mind is a general purpose machine respond to this kind of evidence?

'They would have to say that the mind starts out with a uniform architecture, but due to the statistical contingencies of sensory input from the world, the brain dynamically allocates different areas to different tasks,' Pinker replies. 'But the evidence that a gene deficit can affect a person's ability to conceptualise three-dimensional space is hard to reconcile with the idea that the mind is a homogenous structure.

'The general problem is that nobody has come up with an explicit theory explaining how a learning device with no innate structure could accomplish all the things which a functioning human can accomplish. Whenever anybody tries to implement such a device, they end up building in many kinds of innate architecture. The philosopher W. V. O. Quine made the logical point that no learning system can learn without some kind of innate organisation, because there are always an infinite number of generalisations that are consistent with any set of observations. Even the arch-empiricist B. F. Skinner had to allow for an innate similarity space in order to explain how animals can generalise from stimuli they have been trained on to those they have not been trained on. Similarity is in the eye of the beholder, so, if nothing else, the organism's similarity metric has to be innate. This logical point must always come to the forefront when people talk about how cognitive processes work.'

Evolutionary psychology then is committed to the idea that the brain has a heterogeneous, complex structure. But presumably the idea isn't that there are particular blobs of brain which are identifiable as distinct parts. Rather, we're talking about *functionally* distinct, rather than physically distinct, parts of the brain.

'That's right,' Pinker confirms. 'The analogy is with organs of the body such as blood or the lymphatic system. These don't have dotted lines around them, with one arrow in and one arrow out. Nonetheless, they have characteristic properties which are in the service of particular biological functions. Even the hard disk of a computer is not cleanly divided up into different programs and files: data can be physically implemented in fragments scattered across the disk. Surely the brain is at least that dynamic. The distribution of its functional parts in space may have little causal significance, except perhaps in terms of the speed of connections. So you wouldn't expect the functional units in the brain to be neatly segregated in space. All the brain has to do is to process the information in the right way via the right interconnectivity in the microcircuitry.'

'Of course,' he continues, 'the organs of the brain obviously have a physical implementation, and I would expect this to be found when we understand more about how to parse the brain along functional lines. It will be to do with circuits, though, not big blobs.'

Perhaps the major challenge which is faced by evolutionary psychology is the need to develop techniques by means of which it is possible to identify what the various organs of the brain – as they are manifest in cognitive abilities, behavioural dispositions, and so on – have evolved to accomplish; and indeed whether, in any particular case, they are adaptations at all. Pinker talks about this as the necessity to engage in 'reverse engineering'. Are there any general principles which will result in this being done well?

'Any function that the system has been designed to implement must have a direct or indirect Darwinian payoff,' he answers. 'Therefore, good reverse engineering must show that a part of the mind ultimately has the function of promoting reproduction, normally by the means of achieving sub-goals such as staying alive, not falling off cliffs, making friends and influencing people, understanding the world well enough to manipulate it, finding mates, and so on. Moreover, a putative mechanism for achieving these goals must have an engineering design that is capable of achieving that goal in the kind of environment in which humans evolved. It can't rely for its functioning on aspects of the environment which we know are recent – for example, government, written language or modern medicine.'

'Reverse engineering fails when the reverse-engineer fails to come up with an engineering design that accomplishes a particular adaptive outcome,' Pinker says. 'For example, I'm unimpressed by all the adaptationist accounts of music. The evolutionary functions which have been proposed for music beg the question of how they serve the Darwinian goal of increasing the numbers of copies of the putative genes for music. For example, if someone proposes a hypothesis that the function of music is to bring the community together, the problem is to explain why hearing certain acoustic signals, in particular harmonic and rhythmic relations, should bring a group of organisms together. This is as big a mystery as the question of why people listen to music in the first place. Even if group cohesion is a legitimate Darwinian function, and probably it is, there isn't an independent engineering analysis that leads from the properties of music to group cohesion, unless you already *assume* that the brain is organised in such a way that certain harmonically and rhythmically organised sounds make people want to bond. But, that, of course, is what we need to explain.'

Pinker's scepticism about adaptive explanations of music is interesting because some critics have suggested that evolutionary psychology is in the business of inventing, *ex nihilo*, 'just-so' stories to explain how any cognitive attribute or behavioural disposition was adaptive in our evolutionary past. More specifically, the argument is that we don't know enough about the evolutionary past of humans to specify with any kind of certainty the evolutionary pressures that we have faced. Is this then a fair criticism?

'There surely are problems with *specific* hypotheses and research programmes, but they're not problems in principle for the entire enterprise,' Pinker answers. 'We actually know a lot about the environment in which we evolved, because we know when particular historical developments occurred. For example, we know that 50,000 years ago we didn't have computers, or dishwashers, or written language, or organised cities, or governments, or the rule of law, or modern medicine. Although these are negative statements, they have important implications. For example, the thesis that a taste for revenge is a biological adaptation hinges on the fact that there was no police force or judicial system in our evolutionary past.

'Also, there are certain recurring adaptive problems that all mammals, including humans, face. We know that our ancestors had to attract mates, that they were subject to predation, to parasites, to exposure and to starvation. We know from palaeoanthropology that tools have been part of the human lifestyle for at least two and a half million years; and that meat has been part of the human diet for several million years. So while it is true that we don't know everything, it is just as true that we know a great deal.'

What then is the epistemological status of the more well-established theories or truths of evolutionary psychology? Presumably, we're not ever going to get the kinds of certainty that we get in the hard sciences?

'No, of course not, but that is the wrong comparison,' Pinker replies. 'The temperature at which water boils, for example, is a simple fact about the physical world. But humans are amongst the most complex entities in the Universe, perhaps *the* most complex, and for things as complex as a human you're not going to get laws which are equivalent to the boiling point. The right comparison is not between evolutionary psychology and physics. It's between evolutionary psychology and non-evolutionary psychology.'

Part of the point of this line of questioning has to do with the kind of tactics which people use when they want to undermine Darwinian explanations. To take an extreme example, people who are committed to Biblical creationism will often argue that because nobody has ever observed speciation (which, it should be said, is not true) it remains a hypothesis, rather than a scientific fact. Arguably this makes the issue of the epistemological status of the softer sciences an interesting question.

'It is apples and oranges,' Pinker insists, when I put this point to him. 'If you demand the same confidence level for psychology that you do for physics, or even, for example, the dating of cut-and-dried historical events, you won't do psychology at all.'

But presumably Pinker isn't conceding the point that creationists make about Darwinism: that you can't really be confident about things like speciation?

'Biology is less certain than physics, and psychology is less certain than biology, that's true,' he replies. 'But having said that, we might have a confidence level of 99.999 per cent about Darwinism; it's just not as high as the 99.9999 per cent confidence level, say, that we have with Newton's laws. But it would just be scholastic to argue that because the theory of evolution hasn't been demonstrated with 100 per cent certainty that there are good grounds for doubting it. In the case of evolution, evidence converging from different sources makes the case for Darwinism overwhelming.'

One of the most frequently levelled criticisms of biological explanations of mind and behaviour is that they are determinist; that is, they reduce human beings to the status of automatons, blindly following the dictates of their genes. Does this kind of criticism have anything to it?

'Most people have no idea what they mean when they level the accusation of determinism,' Pinker answers. 'It's a non-specific boo-word, intended to make something seem bad without any content. Nobody believes, for example, that a gene that is said to be "for" a particular behaviour will *always* result in that behaviour. In evolutionary biology, "a gene for x" is shorthand for a gene which, compared to its alternative allele, raises the probability that the organism will manifest a particular behaviour in some environment. The criticism of determinism relies on a literal-minded misunderstanding of a bit of biological shorthand. Nobody has ever proposed that human behaviour is deterministic in the mathematical sense that an outcome occurs with a probability of one.'

It is almost inconceivable that the scientifically informed critics who level the charge of this kind of determinism don't know that they are attacking a straw-man. So what's going on here? Are the criticisms of people like Richard Lewontin and Steven Rose motivated by overriding political concerns?

'In the case of Lewontin and Rose, they are open about the fact that their motivations are largely political,' replies Pinker. 'They say at the outset of their book, *Not In Our Genes*, that they see their critical science as being an integral part of the struggle to create a socialist society, and insist that one cannot separate political motives from scientific theories. They are also so committed to the idea that genetic and evolutionary approaches to mind and behaviour are bankrupt that they refuse to consider reasonable interpretations of such claims, instead criticising the most simplistic interpretations they can arrive at.'

The view which holds that genetic and evolutionary approaches have no significant part to play in explanations of human behaviour is nearly always associated with the idea that the mind is a 'blank slate'; that is, with the idea that the specific content of the mind comes in from the outside. However, it is very difficult to see how this view can be reconciled with, for example, the systematic differences in sexual behaviour which exist between men and women. The evidence shows that across cultures: men tend to want more partners than women; women place more importance on men's financial status than vice versa; men are more concerned with good looks than women; women are more worried about loss of commitment than men; and so on. How can a blank slate view even get started in explaining this kind of thing?

'It is difficult, and in the end I don't think that it can be done,' replies Pinker. 'If you believe that everybody's slate is blank, then by definition the slates of men and women must be the same. Any differences between men and women would have to be caused by factors outside the mind; for example, by the way that boys and girls are brought up, or the way that men and women are rewarded for their behaviour.

'There are then just two main options to explain a systematic sex difference. The first invokes historical accident: perhaps all human cultures are connected to some prehistoric culture which just happened to have that sex difference. The second invokes the constraints that are exerted on the problem space by other causes, which lead to sex differences as an outcome. For example, if men have all the economic power in all societies, then women will value status and earning power in a mate because it is the only means available for them to gain wealth. This is a reasonable hypothesis, which has to be tested. One test is to see whether women's desire for high-status, wealthy partners is lessened in societies where they are wealthier than men. The answer is that it is not. Another test is to see whether in our society, wealthier women have less of a desire for high-status, wealthy partners than less wealthy women. And again, the answer is that they do not.'

In *The Blank Slate*, Pinker argues that one fear associated with biological explanations of mind and behaviour is that if it is shown that there are innate differences between particular groups of people – as, for example, in the case of these sex differences – then, more likely than not, the result will be discrimination and oppression. Is this fear justified?

'One political reassurance of the doctrine of the blank slate comes from the mathematical fact that zero equals zero; if we're all blank slates, we must be equal,' he replies. 'It doesn't follow, though, that if we're not blank slates then there will be significant differences, for example, among ethnic groups, or between sexes. But, as we have seen in the case of women and men, it is certainly possible. The response to this possibility is that we can't legislate empirical facts on the basis of what is morally most comforting. Rather we have to ground our moral convictions on arguments based on identified values. One value holds that individuals should be treated on the basis of his or her merits; that we should not prejudge people because of their ethnicity or sex. Given that value, we can uphold policies of non-discrimination and equality of opportunity regardless of what the facts turn out to be about the average traits of particular groups. We choose to set aside certain actuarial statistics in making decisions about individuals.'

Presumably if we deal with the possibility that particular groups have different average traits by simply denying that it is possible, then if it turns out at some future point that the evidence for the proposition is undeniable, we're in trouble, because there is no further position that we could occupy.

'Absolutely,' agrees Pinker. 'If important moral values hinge on empirical dogma, then you make these values hostages to fortune; the possibility exists that future discoveries might make them obsolete. One needs a firmer foundation than empirical dogma to condemn things which obviously should be condemned.'

There is a second fear associated with biological explanations of mind and behaviour which is perhaps more difficult to deal with than the worry about innate traits and oppression. This has to do with free will and moral responsibility. There are good reasons for us to behave as though we have these things, even if it turns out that in some senses we do not. However, it isn't quite clear what we might do if the situation arises so that we come to understand enough about brain processes to know, for example, that being wired up in a certain kind of way impairs people's ability to make moral judgements. In this situation, wouldn't it be quite difficult to sustain the position which has it that these kinds of people are straightforwardly blameworthy for their actions?

'It complicates the issue, but it doesn't make it intractable, or raise a foundational problem for a materialist analysis of behaviour,' replies Pinker. 'Let's take a clear case: a 5-year-old shoots a playmate. We don't consider the 5-year-old to be blameworthy. Why not? I think it is because the intuition of blameworthiness is closely tied to the intuition of deterrability. There are certain people whose behaviour can be altered by being embedded in a society that has a publicly stated limit to acceptable behaviour and a commitment to punish acts that stray outside those limits. However, we tend to feel that this is not the case for a 5-year-old: the threat of criminal punishment is not factored into the brain system that causes their behaviour. So if the moral goal of a system of responsibility is to deter as much harmful behaviour as possible while avoiding as much suffering as possible – other than that which is justified by the first goal of deterring harmful behaviour – we won't lock up the 5-year-old, because the harm done to the 5-year-old wouldn't reduce the murder rate amongst 5-year-olds generally.

'The hard question is what to do about a person who lacks the neural equipment to respond to the contingencies we call "responsibility". In a sense, we already face that problem with the issue of insanity defences and preventative detention. If you have strong reasons to believe that someone is going to hurt himself or someone else, you can set into motion a complicated legal apparatus that can lock them up, typically in a psychiatric hospital. This may not be an ideal system, but it is a challenge we already deal with; the challenge we might face in the future may be tougher but it is not fundamentally different.'

As a policy to deal with these kinds of issues, this is clear and sensible. But the question remains about whether people who lack the neural equipment to do proper moral reasoning remain *blameworthy*.

'I loathe Mark David Chapman for murdering John Lennon, even though I know he was nutty as a fruitcake,' says Pinker. 'Am I right to hate him? That isn't easy to answer. I think I am, in the sense that if enough people hate him, the certainty that he is hated, to the extent that he can appreciate it, would outweigh the fame and notoriety that was one of the motives for shooting John Lennon. But what happens if I'm wrong? What happens if it doesn't matter how much I hate him – if he's nutty enough to do it, and doesn't care about any consequences? There is no clear answer to that question.'

Is the issue then of free will and moral responsibility more difficult to think through than the worry about innate traits and oppression?

'There may be a component of the problem of consciousness, and a component of the problem of free will, which will never be solved. I suspect that they are not scientific, but rather conceptual, problems. I am in sympathy with people who say that science should ignore the unsolvable aspects of these problems. Nonetheless, being humans, having the minds that we have, there will always be an itch that we cannot scratch. Is a person really blameworthy if there is no contingency of responsibility which will deter him, and hence is blame pointless? Why does it feel like something to have neurons firing in particular parts of the brain in particular patterns? Scientists say it just does. The curious intellect will say, I'm not satisfied. I won't be surprised if this discrepancy is permanent.'

An interesting aside to all this talk about moral responsibility, deterrability and the necessity of punishment is the fact that advocates of the blank slate – and perhaps this is part of its attraction – are required to claim that there is nothing malign in any aspect of the nature of human beings. Is this an idea backed up by the evidence?

'Absolutely not,' insists Pinker. 'First, violence is a human universal, including murder, rape, grievous bodily harm and theft. It is found in *all* cultures. Also, contrary to the belief of many intellectuals, the best ethnographic accounts show that the highest rates of violence are found in pre-state, foraging societies. Rates of homicide and death by warfare in these societies are, by orders of magnitude, higher than in the modern West.

'Even in Western societies, a majority of people harbour violent fantasies about people they don't like, though of course most never act on them. Also, violence appears early in the life of a child – the majority of 2-year-olds kick, bite and scratch. Violence is also common among chimpanzees, our close evolutionary cousins. When you put all this together, it suggests that at least the urge to violence, if not necessarily violence itself, is part of our nature. It's a desire that is present in most people and a behaviour option which we take up easily.'

So from Pinker's point of view the utopian dream that we might one day achieve a society characterised by spontaneous cooperation and an absence of violent tendencies is unrealistic?

'Yes, I think you could say that,' he laughs. 'You have to buy into the myth of the noble savage to believe it is possible, and I don't.'

Despite the advances in biological explanations of human behaviour, Pinker has argued that it is possible that some things are beyond our ability to sort out properly. The *hard problem* of consciousness – how it is that the activity of the brain is accompanied by an inner feel – is, as he has mentioned, one of these things.

'I don't see this possibility as imposing limitations on the science of consciousness,' he tells me. 'Admittedly, a lot of scientists bristle at the idea; they think it is an attempt to wall off some scientific problem by saying that it is futile to study it. But we can learn an enormous amount about the so-called "easy problem" of consciousness, namely the distinction between conscious and unconscious information processing in the brain; and what we can't learn about may not be within the realm of science at all, since it may not have objective empirical consequences. We may simply be bothered by a deep dissatisfaction that we can formulate certain problems without solving them.'

If we were a different kind of being, might we then have a better handle on the problem of consciousness?

'That's an important but almost perversely unanswerable question,' he replies. 'But I suspect we would. There is every reason to believe that certain patterns of neural activity are necessary and sufficient for subjective experience. We just don't know how to bridge the gap. Why should firing neurons *feel* like something? Firing neurons *do* feel like something. I *know* that they do, because I am feeling something right now. But we have no way to bridge that gap. The fact that we are so confident that it is true, yet the problem seems so intractable, suggests that the difficulty lies in the way that the human mind conceptualises the whole problem.'

The term sociobiology was coined by the naturalist Ed Wilson. He now says that the controversy over sociobiology and evolutionary psychology is largely dying out. Is this a view which Pinker shares?

'No,' he replies, 'I am less optimistic. What *has* changed is that the blank slate is no longer an unquestioned dogma, and theories which invoke human nature are no longer taboo. For example, it is not considered politically retrograde and morally suspect to investigate whether sex differences or individual differences have genetic causes. Nonetheless, there are still many people who treat the idea of human nature with fear and loathing.'

3   The Human Brain and Consciousness

In conversation with Susan Greenfield

According to English biologist Thomas Huxley, 'How it is that anything as remarkable as a state of consciousness comes about as a result of irritating nervous tissue is just about as unaccountable as the appearance of Djin when Aladdin rubbed his lamp.' Indeed, it is considered by some to be so unaccountable that we're never going to be able to understand it. Colin McGinn, for example, in noting that the problem 'has stubbornly resisted our best efforts', argues that 'the time has come to admit candidly that we cannot solve the mystery'.

This is not a view which pleases Susan Greenfield, Fullerian Professor of Physiology and Director of the Royal Institution of Great Britain, and Professor of Synaptic Pharmacology at Oxford University. 'It's annoying,' she tells me. 'Can you imagine Christopher Columbus saying "Oh no, we can't sail out of this harbour, we're not going to get anywhere." We'd still be in caves if we thought like that. Also, if we had been told in the middle of the twentieth century that we were going to be carrying around tiny phones by the century's end, or that we were going to have email, we would have found that a strange idea. So I'm optimistic that we can make similar progress with consciousness. If I'm going to throw in the towel, it's going to be at the end of my life, not half way through!'

Greenfield's interest in consciousness is fairly unusual in a scientist.

Francis Crick, for example, has pointed out that only a few years ago it would not have been possible to use the word consciousness in a paper for a journal such as *Nature*, or in a grant application. Things have changed, but consciousness remains a difficult issue for scientists.

'The problem is that subjectivity is central to consciousness,' Greenfield explains. 'Scientific method involves ruthless objectivity. This means that the subjectivity of consciousness is anathema to most scientists because it can't be measured. If you can't measure it, then it is very hard to be objective about it. But the trouble is that if you just ignore subjectivity, then you're ignoring most of what is important about conscious experience. Just because you can document an interesting property in the brain doesn't mean that you've got to grips with consciousness. You find that scientists often engage in a thought process which goes something like this: consciousness is interesting, this property of the brain is interesting, therefore this property must somehow be related to consciousness. But, of course, it isn't straightforward in this way.'

However, although the precise nature of the relationship between brain and conscious experience remains unclear, this does not mean that there is nothing further to be said on the matter. Greenfield's view is that consciousness could be studied as a *correlate* of assemblies of neurons (the nerve cells of the brain), of varying size.

'I like this idea, because it is quantifiable', she says. 'If you look in the brain, what you find are assemblies of neurons working together, and they get bigger and smaller correlated with particular states of consciousness. But the important point is that this is an *index* of consciousness, it is *not* consciousness itself. If you were to put an assembly of neurons in a bucket, you wouldn't get consciousness. So I'm aware that although I might have come up with a very sensitive measure of consciousness, I haven't come up with consciousness itself.

'It is possible to extend this conception to include the *readout* of assemblies of neurons,' Greenfield continues. 'In my view, this readout is a chemical signature which influences the great control systems of the body – the endocrine system, the immune system, and so on. It is also possible to see these control systems as themselves feeding back into the brain via chemical messengers which then influence the size of neuronal assemblies. Although I'm very comfortable with this – the conception of consciousness has been extended to include an iteration of chemicals linking the control systems of the body – it still doesn't *explain* how you get a translation into a subjective sense.'

The puzzle of this translation has been termed the 'hard problem' of consciousness by philosopher David Chalmers. In essence, the problem is that it seems that we could fully understand the functioning of the brain, how it processes and acts upon information, and yet be left with a further question: why doesn't all this go on 'in the dark', free of any inner feel?

One possible answer is that consciousness is an epiphenomenon; that it is just a brute fact about this world that certain kinds of highly organised networks of neurons will result in mental events which have a subjective feel, but that these mental events themselves play no causal role in our behaviour. If this view is right, perhaps there will be no further question to be answered about subjective experience once we have an accurate index of consciousness?

'It seems to me that this kind of argument is equivalent to invoking ghosts,' Greenfield says. 'You can't do anything with it, so it is the last thing you should assume, not the first. It's a bit like saying I've eliminated burglars, I've eliminated creaky floors, therefore it's got to be a poltergeist. Also, you could never know that consciousness is irreducible in this kind of way. There are further questions to be answered. Consciousness varies in degree, for example, and I want to know why, and how.'

But given that Greenfield is offering an account of consciousness which references physical entities – assemblies of neurons – how does she see us as escaping the causal closure of the physical world (i.e. the fact that physical events seem only to have physical causes)? If we can't escape it, then aren't we just sophisticated robots, the product of the physical processes which occur in our brain?

'Well, these are very interesting issues,' replies Greenfield. 'But, as John Searle points out, if we go into a restaurant to order a hamburger, we don't worry about whether it is our genes, for example, driving our behaviour; the phenomenology is that we act freely. Of course, we can deconstruct this behaviour in terms of hormone levels, chemicals and active brain regions, or, in a more chronic way, in terms of genes, and so on. This *is* interesting because it brings to the fore issues to do with accountability. If somebody can say, for example, that they attacked a stewardess on a flight because the alcohol they had consumed had reacted badly with their medication, why can't we all have recourse to these kinds of explanations? The issue is how to balance this up against the concept of free will.

'My view is that we have to live with both sets of concepts. Think about a depressed person, for example. It doesn't help if I say to them, "Oh, you're not really depressed, it's just that your serotonin levels are low." They're still depressed. Just because we might be able to explain the biological correlates of our feelings and thoughts doesn't make those feelings or thoughts any less real. If you make certain choices, and I'm able to tell you their biological correlates, they are still *just* correlates. They shouldn't be used to diminish the sense that people are individuals and, therefore, accountable for their actions.'

However, whilst it might be true that our *sense* of being individuals is not diminished in this way, it isn't clear that we are freely acting in quite the way that many people might think. For example, if it is the case that the person who ends up murdering someone in a park was always going to do so because of some complex interaction between genes, brain and environment, then surely, in a very real sense, they are not accountable?

'But this is such a complex situation that you don't really get this kind of determinism,' Greenfield replies. 'It is much more analogous to the chaos theory model. Our behaviour is such a complex function of the interaction between genes, brain, events in our lifetime, the environment, and so on, that although it might be possible *post hoc* to deconstruct certain dominant themes, it doesn't mean that these were always going to hold sway. We won't ever be able to come up with simple explanations as to why people behaved the way that they did. And it is in this complexity that our individuality and our accountability reside.'

These kinds of issues come to the fore in the case of humans because we are reflective, self-conscious beings. For Greenfield, self-consciousness is different from consciousness, and is linked to the development of the mind.

'My view is that the mind is the personalisation of the brain,' she explains. 'I certainly don't think that it is some kind of ephemeral alternative to the squalor of the biological brain. Mind has a physical basis. We know that a person is born with all the brain cells that they'll ever have, but it is the growth of neuronal connections which accounts for the growth of the brain after birth. These connections are highly sensitive to events in the environment. We are born into buzzing and confusion, but as we get older our experiences coalesce into faces and objects, which acquire labels and which gain meaning as they feature in different events in our lives. It's this personalisation of neuronal connections which I would call the mind. And it's equivalent to self-consciousness. When we talk about "losing our mind" or being "out of our mind", what we're really talking about is not being self-conscious.

I think that this shows that it is absolutely wrong to conflate mind and consciousness. The two are separate; it is quite possible to lose one's mind, but to remain conscious.'

One of the consequences of separating out consciousness and the mind in this way is that it lets in emotion as the most basic form of consciousness.

'Emotion is what the one-day-old baby experiences,' confirms Greenfield. 'It's what the purring cat and the dog wagging its tail experience. It's what we get back to when we don't access our neuronal connections, and therefore don't make use of our learning and memory. We go back to living in the moment. Children tend to experience the extremes of emotion more readily than adults. They have fewer neuronal connections available to them. We know that's the case in a young brain. And because there are fewer connections, the transient assemblies of neurons cannot be as big, and therefore consciousness is more raw, it's not fully self-conscious. A young child does not access a past or a future, they're evaluating the world pretty much in purely sensory terms, rather than seeing it in terms of the context of what has happened before.'

There is, however, something slightly counter-intuitive about this idea, in that Greenfield seems to be suggesting that the intensity of experience varies *inversely* with the size of the assemblies of neurons. The smaller the assemblies, the greater the emotion.

'Yes, but you've got to bear in mind that the intensity of experience is not just about neuronal assemblies,' replies Greenfield, when I put this to her. 'It's also about how the hormones and chemical composition of your body are affected. And it isn't so much that children have more intense experiences – because it would be difficult to know such a thing – it's rather that their experiences aren't held in check like in an adult mind. If a child drops an ice-cream on the pavement, it's a real disaster, but adults will reason that they are able to buy another one, that people are dying in the Middle East, so it's not that important. They are able to temper their reactions. But if a person is a schizophrenic, or if they're dreaming, or if they're a baby, then they don't have these checks and balances, so the emotion is more raw.'

'The key point is that connections in the brain function to contextualise experience,' Greenfield emphasises. 'A small child will be frightened by someone dressed in a sheet pretending to be a ghost because there is nothing to rationalise the experience. But an adult will know it is just a friend in a sheet. This isn't entirely a good thing – we miss out on wonderful childlike reactions – but it does mean that we protect ourselves in that the world makes sense to us and we're not at its mercy like we are as children.'

Is there, I ask, a more general phenomenological correlate here? Do adults have a greater sense of themselves as individuals if they have more extensive neural connections?

'Yes, I think they do,' answers Greenfield. 'But it is necessary to be a bit wary when talking about individual differences. The human brain encompasses such a wide range of levels of consciousness that you've got to be a bit careful before concluding that someone like Van Gogh had a deeper sense of consciousness than you or I, even if we suspect that it might be true.'

Greenfield's view then is that the mind is constituted by networks of neuronal connections which are built as a result of personal experience and learning. But how is it that we're conscious of whatever it is that we're conscious of at any particular moment? What's going on at the level of neurons?

'I think of this in terms of an analogy with a stone in a puddle,' she responds. 'There are all these latent circuits in the brain which are not active. To become active, they must be triggered by something. In our waking moments, it is normally by interaction with the outside world. Although sometimes we might become immersed in a fantasy or a daydream, most of the time when we are awake, we're interacting. So, for example, if someone walks into a room, then that will be at the centre of our consciousness because it is a powerful enough stimulus to recruit a large neuronal assembly so that we become conscious of it.'

'The centre of the consciousness of children,' Greenfield continues, 'tends to be determined by straightforward sensory stimulation; for example, as when an alarm bell goes off. Children ricochet from stimulation to stimulation. But as we get older, this kind of pure sensory stimulation is replaced by things to do with *significance*. So, for instance, if your lover comes into a room, in physical terms he won't have any particular sensory impact, nobody else will notice him, but because he is linked with so many of your memories and associations, he will register in your consciousness. As we get older, then, we substitute large networks of connections for strong activation of brain cells, and these have the same power in recruiting neuronal assemblies.'

So what about the common situation where a thought or memory pops unexpectedly into our heads? What happens so that we suddenly become conscious of it?

'We know that there are lots of subconscious things going on in the brain all the time,' Greenfield answers. 'What makes them conscious is that they become co-opted by a large neuronal assembly. A bit like the way blobs of mercury might gel. I think this is a large part of what constitutes creativity. It often seems like suddenly we have a really good idea, but presumably the brain has been trying out all kinds of ideas for a long time, and we become conscious of the good idea only when it becomes co-opted into a neuronal assembly.'

Greenfield's idea then is that there are different levels of consciousness. This seems highly plausible. We have all had, for example, the experience of not being fully conscious on first waking in the morning. So why have people been reluctant to accept it?

'Well, I should say that I'm not sure that I'm the first person to have said this kind of thing,' she replies. 'But I am the first person to say it so emphatically. The fundamental thing, and where people have had the stumbling block, is the idea that consciousness is not all or nothing. Once you begin to think about this, you realise that it actually solves lots of problems. It gets around the difficulties of thinking about foetal consciousness, animal consciousness, and intuitively it seems to make sense that consciousness varies in degree in line with our physiologies.

'If you buy into this idea, then everything is changed. It means that you can look for something in the brain which varies in degree, rather than looking for some magic brain region. I came to this conclusion about consciousness by studying unconsciousness. There are different degrees of unconsciousness, different degrees of anaesthesia and sleep. It seems to follow that if you have degrees of unconsciousness, then you must have degrees of consciousness. As soon as you think this, then you turn something qualitative and ineffable into something quantitative and scientific.'

So self-consciousness, the fully developed mind, if you like, is at the extreme end of a continuum of consciousness?

'Yes, and the kind of consciousness which you get when you're playing a fast-paced sport, for example, is towards the other end; you're in the moment, it's just stimulus and response. What's interesting is that we seek out these kinds of experiences, indeed we pay money to have them, but if you have them all the time, then you're unfulfilled. The human condition requires that we achieve a balance between, on the one hand, letting ourselves go and, on the other, having some kind of fulfilment and satisfaction of the private ego.'

There is an interesting link here to some of Greenfield's thoughts about depression. In her book *The Private Life of the Brain*, she argues that depression will often be associated with extensive networks of neuronal connections and an over-concern with the private ego. Presumably then depressives have a strong sense of themselves?

'Yes, and, at least partly, that's why they're depressed,' Greenfield agrees. 'It would be lovely if depressives could lose themselves in dancing or playing squash, but they can't. They sit around with a persistent obsession with the self, feeling cut off from the outside world. But precisely what causes depression is a very complicated story. One of the fascinating things about the brain is that a cognitive stimulus – for example, whispering in a person's ear – can produce a whole series of physical changes. This puts one in a chicken and egg situation with respect to depression. Either there is a change in brain chemistry which results in depression; or some cognitive or physical event leads to a physical change, which then results in depression. But whichever it is, the effects are the same. Depressives are stuck in a world of over-extensive neuronal connections.'

Are there any therapeutic consequences of this way of seeing depression?

'I have the controversial view that drugs aren't the answer,' Greenfield replies. 'As a short-term measure, if someone is in great physical or mental pain, it would be inhuman not to help them with drug therapy. But then all you're doing is giving people chemical oblivion, you're not solving the underlying problems. If you give someone Prozac, it's not going to make any of the cognitive stuff go away. If a person has been dumped or their father has died, these things are going to be true after the Prozac wears off.

'I think that cognitive therapy is probably best. If you can help someone to see something in a different way, and get them to absorb it, accept it and live with it, then whilst it won't make their problems go away, it will make them more manageable; and I think that the way that it does this is by reordering the connections in the brain, which is much better than simply treating someone with Prozac.'

Greenfield's view has is it that emotion and mind are the opposite ends on a continuum: the purer an experience of emotion, the more that we 'lose our minds'. This view dovetails in an interesting way with her attitude towards drug taking. Greenfield has argued that the impact of a drug like ecstasy is to render the mind of a person like that of a child. Is this the reason that she is on record as saying that drug taking should remain criminalised?

'There are several things going on here,' she answers. 'I go along with the idea that we all need times of ecstasy, of being out of our minds. And these can be achieved with orgasm, or fast-paced sports, or dancing or fine wine. We all seek moments like these, but we want to be doing other things as well. However, if we have a drug culture, then the balance shifts so that people are mainly seeking out these kinds of experiences, and then they will lead relatively unfulfilled lives.

'Also, I think that if people see drugs as a solution to their problems then this diminishes their humanity. Being human is about problem solving, evaluating things, seeing things as a challenge and overcoming them. By simply reaching for drugs we do ourselves down, we don't do ourselves justice.

'So there is a moral stance here to do with what it is to be a human being. But there is also medical evidence that there may well be long-term damage in taking drugs. I know that drug taking doesn't kill people. But if by damage we're talking about changing the personality of a person, about altering their attention span, about affecting their potential, then drugs *do* harm people.'

Although Greenfield is keen to stress that from time to time humans need to escape the associations and connections which constitute their private selves or private ego, she also thinks that there are dangers in not having access at all to a fully developed private ego. Particularly, she is interested in the notion of a public ego, and how this might fill the void left by a diminished private ego.

'What triggered my thinking about this is a book by Sebastian Haffner called *Defying Hitler*,' she explains. 'He sets out to explain why Hitler gained such purchase in Germany. He argues that during the First World War, German people began living collectively; they started to develop a public ego. After the war, they had no private life, no private ego or sense of personal fulfilment, and what took their place was sport. This became the thing which everybody could talk about and enjoy a shared identity around. Of course, then what happened was that Hitler came along with an even stronger, more romantic narrative, and he took over. People often make the mistake of thinking that things like the Nuremburg rallies were a return to an animalist, reptilian past, but it wasn't like that. Nazism had a very clear narrative, it provided a very clear identity, but it was a *shared* identity – you were part of a nation with a story and a destiny.

'There's a similarity here with groups like Al-Qaida; there's a romantic dimension to Bin Laden's philosophy. His world is divided up into good guys and bad guys. He talks about gaining revenge for the crusades, and he articulates a romance of the downtrodden.'

One significant aspect of this argument is that it suggests that there is something valuable about individualism and the private ego; that they are an antidote to the temptations of the more extreme mass political movements. Perhaps then there is an interesting reversal here of the traditional left-wing argument that the materialism of capitalism is a bad thing because it encourages individualism?

'Yes, there is,' Greenfield confirms. 'As Bertrand Russell pointed out, collective passions are dangerous. One of the best ways of suppressing them is by encouraging the private self. You can convince someone that they're an individual by offering them choices – for example, this settee or that settee – and this was one of the very interesting things about the first half of the twentieth century: it cultivated a concept of the self which was very new. For the first time, people could consume what they thought that they wanted, and in doing so they were able to develop and express a private self. By doing this then, capitalism hinders people from developing a strong collective narrative. I think there is a conflict in society between a public and private ego. And a public ego is something which is inherently dangerous, so I don't mind a bit of materialism if it means that people think that they are living private lives.'

Presumably this feeds back into Greenfield's views about drugs, especially any drug – for example, ecstasy – which might encourage people to think they are living shared lives?

'Absolutely,' says Greenfield. 'It is very easy to imagine some kind of drug of collective experience. Just think of the *Brave New World* scenario.'

As a working scientist, Greenfield spends much of her time looking at the neuronal mechanisms which are common to both Alzheimer's and Parkinson's disease, with the aim of developing ways of preventing the neuronal death which occurs with these illnesses. But is she, I ask, also able to spend time researching and testing her ideas about consciousness and its relationship to transient neuronal assemblies?

'There is a lovely phrase that working on consciousness is for a neuroscientist a CLM, a Career Limiting Move,' she answers. 'There is no way that the public sector charities and research foundations will fund this kind of research. It's bad enough trying to get funding for the more pedestrian stuff, but with something so seemingly airy-fairy and mystical as consciousness it is almost impossible. However, that being said, I do have a little money now for some imaging equipment, and, in a small way, I'm going to start looking at the mechanics of the neuronal assemblies in brain slices. I care passionately that my theories are testable. Much more so than about whether they are right or wrong. Being testable is what counts, it is what makes it science.'

4 Science and the Human Animal

In conversation with Kenan Malik

In his book *Man, Beast and Zombie*, Kenan Malik argues that human beings are quite unlike any other organism in the natural world. We have a dual nature. We are evolved, biological creatures, with an evolutionary past, and in this sense we are simply objects in nature. But we also have self-consciousness, agency and the capacity for rationality, and as a result we alone in the natural world are able to transcend our evolutionary heritage and to transform ourselves and the world in which we live. Science, though, is taking its time in getting to grips with this dual nature of human beings.

'Most people will accept that human beings are animals, beings with evolved brains and minds,' Malik tells me. 'Most people will also accept that humans are in a certain way distinct from all other animals. The important issue then is about how we understand the relationship between these two aspects of our humanity. Or more particularly, how we understand, on the one hand, the continuity of humanity with the rest of nature, and, on the other, the distinctive aspects of humanity. What is key is whether the distinctive aspects can be understood in the same terms, using the same tools, as we understand those aspects of our humanity which are continuous with nature.'

In getting to grips with this dual nature, Malik makes use of a distinction between human beings as objects and human beings as subjects.

'It's a critical distinction,' he says. 'The questions to ask are: What kind of beings are humans? Are they the kind of beings which can be fully encompassed within the domain of natural science? The paradox of natural science is that its success and understanding of nature places constraints on its understanding of human nature. The pre-scientific world was full of magic, purpose, desire, and so on. The scientific revolution "disenchanted" this world, to use Max Weber's phrase. It expunged magic and purpose, transforming nature into an inert, mindless entity. At the heart of scientific methodology, then, is the view of nature, and of natural organisms, as machines; not because ants or apes are inanimate, or because they work like watches or TVs, but because, like all machines, they lack self-consciousness and will. It is this mechanistic view that has made science so successful over the past half millennium, enabling it both to explain and to exploit nature.

'But humans are not disenchanted creatures in this way. We possess, or at least we think we possess, purpose, agency, desire, will, and so on. Humans are biological beings under the purview of biological and physical laws. But we are also conscious, self-reflexive agents, with the capacity to bend the effects of biological and physical laws. The very development of science expresses this paradox of being human, because in order to understand nature objectively, it is necessary to make a distinction between nature which is the object being studied and humanity which is the subject doing the studying.

'When you're looking at "external nature" this distinction is relatively easy to make. But when you're studying human beings it is not so easy, because human beings are, at one and the same time, the objects of scientific study and the subjects doing the studying. In a sense, then, in order to have a scientific view of nature, we have to be outside of it, to look down upon it, because if we did not, then we would not be able to understand it objectively. So our very capacity to do science, our very capacity to study nature objectively, reveals paradoxically the sense in which we are not simply immanent in nature, but also in a certain way transcendent to it.'

According to Malik, one of the consequences of the dual nature of human beings is that the reductionist project in science – where reductionism is the idea that it is possible to explain a particular phenomenon in terms of its most fundamental constituent parts – will only result in a limited understanding of the human mind; that explanations of human behaviour and actions should not be couched solely at the level of neurons firing, and so on. What then are the limits of reductionism for understanding human beings?

'Well, it is certainly *possible* to have a reductionist explanation of how the mind works,' Malik replies. 'And we could extend this to give say a reductionist explanation of this conversation. But, in fact, this would tell us very little about the conversation, and what we're discussing, because humans do not operate solely at the mechanistic level. In other words, reductionist explanations are perfectly possible, but in many contexts they are not very useful.

'Let me give you an illustration. Suppose one morning I run out of the house and I kill a passer-by. In a subsequent trial, I might use one of two defences. I could claim that I had been suffering from an inoperable brain tumour which had made me irrationally violent. Or I could say that the night preceding the murder I had had a conversation with a brilliant, but evil, existentialist philosopher, who had convinced me that the only way that I could express my individuality was to murder somebody. In the first case, if the story was true, most people would say that I was not guilty of murder; that I was acting in some fashion over which I had no control. In the second case, however, most people would say that I was guilty of murder. But actually, in *both* cases there was some cause of my action, either a brain tumour or an evil existentialist.

It is just that people will normally distinguish between my acting as an object, as in the first case, and my acting as a subject, as in the second.

'However, it would actually be quite possible to couch the evil existentialist explanation in mechanistic terms; that is, to treat my actions as if I was just an object. I could say that the reason that I committed the murder was because certain neurons fired, in a particular order, in the brain. This is a perfectly valid explanation, in the sense that had those neurons not fired, then I would not have committed the murder. However, in this context, such an explanation is not very useful, since it fails to recognise that humans also act because of conscious reasons, and not simply because of physical causes. Reductionist explanations by and large suffice for non-human animals, but for humans, they do not.'

'I'm not suggesting that reductionist explanations are invalid or don't work or don't help elucidate important aspects of the workings of the human mind,' Malik says. 'All I'm suggesting is that they are insufficient as explanations for how humans work. It's worth adding, I think, that in criticising a mechanistic view of human beings, I am not reaching for some kind of supernatural explanation. The distinction I am drawing is between a *mechanistic*, a *mysterian* and a *materialist* view of the world. A mechanistic view sees human beings largely as objects through which nature acts. A mysterian view suggests that there are aspects of human existence not knowable to mere mortals. A materialist view, on the other hand, understands human beings without resort to mystical explanations. But it also sees humans as exceptional because humans, unlike any other beings, possess consciousness and agency. And understanding human consciousness and agency requires us to understand humans not just as natural, but also as historical and social, beings.'

The claim here then is that it is only human beings who have the capacity for self-consciousness, agency, and so on; that humans act for reasons, but non-human animals do not. It is fairly uncontentious to claim that a stag beetle doesn't have these capacities, but it isn't so clear that our close relatives in the animal kingdom do not. So, for example, there is an argument about whether or not the great apes are self-aware, have a moral sense, and so on. Is Malik's view that even our closest relatives in the natural world are completely lacking in these kinds of characteristics?

'There is considerable debate about whether our closest relatives have the capacity for language, morality, culture, tool making, and so on,' he replies. 'I think a useful way of approaching this debate is to examine what we mean by culture when we talk about it in relation to chimps, and what we mean by it when we talk about it in relation to humans, because I think, in fact, we are talking about different phenomena.

'Most primatologists would say that culture is about the acquisition of habits, and you find that there are groups of chimps who are able to perform actions which other groups of chimps are not – for example, some are able to crack open palm nuts using two stones as a hammer and anvil, which is an acquired habit. By this definition, chimps have culture, and there are forty-odd different habits which have been observed in the wild in different groups of chimps.

'But when you talk about human culture you're not simply talking about the acquisition of habits from others. You're talking about our capacity to transform ourselves and our world through a process we call history. It is about 6 million years since the evolutionary lines of chimps and humans diverged. In that time, both chimps and humans have evolved. But, in a broad sense, give or take the capacity to crack open a few palm nuts or to hunt termites with a stick, the behaviour and lifestyles of chimps are more or less the same as they were 6 million years ago. This is clearly not the case with humans. Over the last sixty or seventy thousand years humans have become a very different kind of being. Humans have discovered how to learn from previous generations, to improve upon our work, and to establish a momentum to human life and culture that has taken us from cave art to quantum physics and the conquest of space. It is this capacity for constant innovation that distinguishes humans from all other animals. There is a fundamental distinction between a process by which certain chimpanzees have learnt to crack open palm nuts using two stones as hammer and anvil, and a process through which humans have created the Industrial Revolution, unravelled the secrets of their own genome, developed the concept of universal rights – and come to debate the distinction between humans and other animals.

'All animals have an evolutionary past. Only humans make history. This distinction is critical. So when we talk about non-human animals possessing culture or morality or language capacity, we need to define what we mean by these – and not to assume that we mean the same thing when we talk about chimp culture as when we talk about human culture.'

According to Malik, language plays a crucial role in facilitating the self-consciousness, rationality and agency of human beings. Indeed, he makes use of a Wittgenstein-inspired argument in order to show that these kinds of things are, in a certain sense, dependent upon language, and also to argue that they have social and public aspects.

'Very simply put, my claim is that meaning is social,' he tells me. 'If you are a solitary creature, you might experience redness or pain or a whole host of other things, but attributing meaning to all those things only comes about through our existence as social beings, because meaning derives from social existence. The contents of my inner world mean something to me, in part at least, in so far as they mean something to others. I can make sense of my self only in so far as I live in, and relate to, a community of thinking, feeling, talking beings. Language is critical to this, in part because it underpins our capacity to be social, and therefore it plays a role in allowing us to attribute meaning to our inner feelings or thoughts.'

The argument that the meanings which one might attribute to something like one's own pain are mediated by language and its attendant public and social aspects is relatively uncontentious. What seems less certain is that the actual *experience* of pain might be transformed in the same kind of way. Is this what Malik thinks occurs?

'There are instances of children who from birth have been deprived of social contact and language, and the way that they understand the world, the concepts they form about the world, are very different from those of other children their age,' he replies. 'Concepts of time, place and space, for example, are either absent or highly impoverished in their view of the world. So yes, I do think that social interaction and language play a very important part in mediating our experiences – and in allowing us to make sense of our experiences.'

But there is a difference between arguing that culture and language are integral to the way in which one might understand something like time, for example, and arguing that they are central to how one might *experience* pain or the flight-or-fight mechanism. If one *reflects* upon pain or the fear we might feel when confronted by a threatening situation, then, of course, language and culture have an impact, but in terms of how the experience actually *feels*, it isn't as clear that they do.

'Well, nobody really knows how language affects experience at that level,' Malik replies, when I put this point to him. 'But I think that what one can say is that without language, one cannot experience these kinds of things in the context of social meaning. In humans, our existence as symbolic, social beings seems to transform even such a basic physiological response as pain. One does not have to be a Baron Masoch to recognise that the experience of pain can sometimes be pleasurable, that sometimes we may seek pain as part of sexual or other forms of gratification. In other words, even a response such as pain can be mediated through language, culture and social conventions, such that the human response to pain becomes different from that of other animals. The ways in which we reflect upon our experiences are clearly mediated through language and culture. But the experience itself – how it *feels* – should not be seen simply as a natural, unmediated physiological response. Experiences too are socially mediated.

'Descartes argued that knowledge of one's mind is the starting point for knowledge of other minds. I'd suggest that without knowledge of other minds, it is impossible to have knowledge of our own. Far from inferring other humans' experiences from our own, we can only truly know what goes on inside our heads by relating to other humans. It is only because we live not as individuals, but within a social community, and moreover within a community bound together by language, that we can make sense of our inner thoughts or feelings.'

Part of the significance of this argument has to do with whether non-human animals, which do not have language, have the same kinds of experiences that we have. Does Malik think that primates, for example, experience pain?

'From a scientific perspective, there is no way we can answer this question one way or other,' he replies. 'Most of the claims about animal experience come from the animal welfare movement, and it has popularised concepts such as "animal suffering" and "animal pain". Most of us, if we see an animal, or at least a vertebrate, that is injured, or placed in what seems to be an unpleasant situation like a small cage, assume that they are suffering or in pain. The feeling is spontaneous, instinctive. Anthropomorphism seems to be part of our nature. We appear to be designed to read other people's minds – and to assume that other animals have minds as we do.

'But there is no scientific evidence that they do. Even Marian Stamp Dawkins, the zoologist who has made a case for the idea of animal suffering, accepts that "no amount of measurements can tell us what animals are experiencing". She argues that the conclusion that animals experience suffering in ways similar to ourselves has to be based on an analogy from our own feelings. Animals writhe, cry out, seem distressed, as humans do when we are in pain. So, it seems reasonable to assume that animals can also feel pain.

'However, animals do many things similar to humans but not necessarily for the same reason. Ants "enslave" other ants, ducks "rape", and so on. But there is nothing reasonable about drawing analogies between slavery in ants and human slavery, or duck rape and human rape. Animal behaviour, in other words, does not always provide a good analogy for human behaviour. In any case, behaviour is not necessarily a good indicator of sensations, even in humans. One study has revealed that among patients admitted to hospital with severe injuries, 40 per cent noticed no pain at the time of the injury of which they were fully aware, 40 per cent had more pain than expected, and only 20 per cent reported the expected pain. Another demonstrated that people given anaesthetic while a tooth is extracted may writhe, jump and cry out as if in pain, yet are so unaware that a tooth has been extracted that they frequently ask when it's going to happen even after the tooth has come out.

'The scientific problem of understanding animal experience is made worse by the fact that concepts such as "sentience", "consciousness" and "self-consciousness" are bandied around without much thought to their meaning. What is the distinction between sentience and consciousness? Is it possible to be conscious without being self-conscious? If not, what does this say about the experiences of non-human animals? Such questions are all too rarely asked.'

Malik's arguments are predicated on a thorough-going humanism. One of the criticisms made of humanists is that they can adopt an unwarranted, purely instrumental stance towards the world; that is, they see the world, and all that is in it, as being there to be transformed according to the will of human beings. Is this a legitimate worry? For example, if we accept the argument that non-human animals lack any real sense of their selves does this mean that there is no reason to accord them the kinds of rights that arguably seem to depend upon self-consciousness and agency?

'Obviously, we're quite capable of treating our natural environment instrumentally, and transforming it, and I don't see anything wrong with this,' Malik responds. 'Indeed, the whole process of human development has been about transforming the environment for our benefit. If you think that this is a problem, then you're going to have a problem with agriculture – and indeed, with just about everything else we have done in the last 20 to 30,000 years.

'Do I think animals should have rights? No, I don't. Rights are an expression of our existence as conscious agents, capable of taking responsibility for our actions. Just as we don't put chimps on trial for murder, nor do we accord them rights. It would be absurd to accord them rights, and anyway in a sense we would not be according them rights at all – in order to have a right, you must be able to assert that right, but chimps cannot do so. You might say that they have rights, but in fact their rights must be invested in human beings who in effect act on their behalf. So even if we wished to give rights to apes it would not be possible without distorting the very meaning of "rights".

'Do I think animal experimentation is wrong? No, I don't. I think that the well-being of human beings who benefit from all kinds of medical and scientific advances which are the outcome of medical research is far more important than the welfare of the animals upon which experiments are performed. It seems to me that those people who argue against animal experimentation have a deeply anti-humanist view in that they do not recognise the importance or the usefulness of the kinds of medical advances that do come from animal experimentation.'

If animals are just zombies, then this view is perfectly understandable. Would it be different, I wonder, if Malik had a different view about the levels of self-consciousness in non-human animals? It is possible that there might be a worry with primates, for example. Even accepting arguments about the importance of language for the development of a sense of self and in mediating experience, it does seem that there is the possibility that primates have sufficiently developed brains that even if they don't have our kinds of self-consciousness and experiences, they are nevertheless self-conscious in some kind of way – and that this renders experimentation on primates morally problematic.

'I'm not sure what "self-consciousness in some kind of way" means,' he replies. 'The main evidence that non-human animals possess self-awareness comes from the mirror experiment. A mark is made on a chimp's face while it is anaesthetised. When it subsequently sees itself – and the mark – in a mirror, it recognises that the mark is on its face. Monkeys, on the other hand, don't recognise that the mark is on them. Gordon Gallup, who first demonstrated this, argues that this shows that chimps possess self-awareness.

'But our conscious selves are not in our bodies. We do not see our conscious selves in the mirror. Autistic children, who are unable to attribute mental states to others or to think of how they appear to others, are able to use a mirror to inspect their own bodies at the same age as normal children. Many animals, from baboons to elephants, are able to use mirrors or closed-circuit television to guide their hand or limb – or in the case of the elephant, trunk – movement. Self-recognition, therefore, is not the same as self-awareness. There are many other explanations for the chimp behaviour apart from self-awareness. For instance, Julian Jaynes suggested that the chimp, unlike the monkey, may have learnt a point-to-point relation between the mirror image and its own body, and hence is able to use the mirror as an optical tool for manipulating its body. The mirror test may indicate bodily awareness or representation rather than the conceptualisation of a self.

'Some argue that since we cannot be sure whether or not animals, particularly primates, possess self-awareness, so the morally correct policy is to assume that they do. I disagree. If as scientists we assume that primates have self-awareness even though we have no real evidence that they do, this can do great harm to science. And what is moral about banning animal experiments that might save, or improve, the lives of thousands, perhaps millions, of human beings, on the grounds that primates may possess self-awareness? As it happens, even if the great apes possess the degree of self-awareness that some believe they do, I would have no problem with experimentation that is important for medical or scientific advances.'

It seems then that Malik's argument is predicated on a fairly straightforward claim about what it takes for something to be a moral subject – it requires things like sentience, rationality, self-consciousness, and he doesn't think that non-human animals have these to any significant degree.

'Yes,' he agrees. 'Moral rights are a product of our existence as subjects. Non-human animals are not subjects.'

Perhaps the central theme running through all these arguments is the notion that human beings are in some sense able to transcend their character as natural, biological organisms, and that it is this ability which marks them out as an animal apart from all others. But there is, of course, research which seems to show that we are thoroughly enmeshed in the physical world, and indeed which suggests that things like agency and consciousness might, in some senses at least, be illusory. Benjamin Libet's work, in the 1960s, on *readiness potential* is a case in point.

A readiness potential is an electrical change in the brain that precedes a conscious human act – such as waggling a finger. Libet found that if volunteers are asked to waggle their finger within a 30 second time frame, the readiness potential that accompanies the waggling begins some 300 to 400 milliseconds *before* the human subject reports that they have become aware of their intention to waggle the finger. This is disturbing, because, as Libet put it, the 'initiation of the freely voluntary act appears to begin in the brain unconsciously, well before the person consciously knows he wants to act!' The threat to Malik's argument here is clear. If our conscious acts are unconsciously initiated, then what of free will and agency? Perhaps we *are* just sophisticated machines after all.

'Actually, I think Libet's work, though interesting, is probably irrelevant to the debate about human agency,' says Malik. 'If you're a materialist, then you're going to accept that behaviour, at one level, is caused by the brain. Therefore, there is nothing startling that Libet shows. Unless you want to believe that behaviour happens without causation, and that free will can only exist in an undetermined universe, then I don't think, in terms of this debate, Libet's results undermine the arguments for agency.

'I agree with Dan Dennett that free will and agency can only exist in a determined universe, because in an undetermined universe what you have is not freedom, but randomness. The problem is simply that we don't yet have a conceptual framework which allows us to think about human beings as both determined and free at the same time. The real task is to create the framework which will allow us to view ourselves in both these ways simultaneously.'

But if one accepts the radical conclusion of Libet's work, that somehow consciousness surfs the wave of our brain's physiology, it does seem that reductionist, causal explanations of behaviour – that is, mechanistic explanations – are perfectly valid. It seems that what actually drives our behaviour is just the unfolding of a causal process which occurs at levels below things like intentions, motives and meaning.

'Certainly, there are people who will argue this position, who want to see us simply as biological objects,' agrees Malik. 'The problem with such arguments is that, by their own criteria, they provide us with no reason for believing in them. If "intentions, motives and meaning" are merely epiphenomena, so too is truth. That's because "truth" has no meaning in a world composed simply of objects. As the anti-humanist philosopher John Gray has put it, "in the struggle for life, a taste for truth is a luxury", even a "disability". From an evolutionary point of view, truth is contingent. Darwinian processes are driven by the need, not to ascertain truth, but to survive and reproduce. Indeed, the argument that consciousness and agency are illusions designed by natural selection relies on the idea that evolution can select for untruths about the world because such untruths aid survival. So, if we were simply biological objects, not just consciousness and agency, but truth and reason would disappear, too. In which case we couldn't place any trust in the claim that we were just biological objects. Far from science revealing humans to be beings without consciousness and agency, we are able to do science only because of our ability to act as subjects, rather than just as objects.

'Certainly we need a materialist account of our existence, but a materialist account that doesn't make consciousness and agency disappear in a puff of mechanistic smoke. The problem in trying to understand humans both as biological beings *and* as agents with a certain degree of freedom is that currently we have no conceptual framework to reconcile these two views of ourselves within a materialist world view. I have no idea how to solve this conundrum, but I don't think that anyone else does either. However, unless we start asking the right questions about these issues, then we're never going to get to the right answers.'

5 Cybernetics and a Post-Human Future

In conversation with Kevin Warwick

Kevin Warwick, Professor of Cybernetics at the University of Reading, is a controversial figure. His academic credentials are first rate, including a professorship at the age of 34 and more than 300 published papers. Yet when it was announced that he was to give the first Royal Institution Christmas Lecture of the new millennium, Professor Susan Greenfield, the Director of the Royal Institution, was moved to declare her support for Warwick, after his appointment was criticised in certain sectors of the academic community. And despite the fact that he is the recipient of awards such as MIT's 'Future of Health Technology Award', *The Register*, a high-profile technology web site, has run a series of articles which are more than a little critical of his work.

The proximate cause of the controversy which Warwick attracts seems to be an unhappiness about the high media profile which he enjoys. He has, for example, written a number of popular books on issues to do with robotics and cybernetics, including *In the Mind of the Machine*, where he argues that once machines become more intelligent than humans, and capable of autonomous learning, they will come to dominate humankind. This claim has provoked strong critical reaction, and has been called variously *silly, outlandish* and *ludicrously quaint*. However, Warwick is not alone in thinking this a real possibility. For example, Steven Hawking, Hugo de Garis, Bill Joy and Hans Moravec have all argued that it is not inconceivable that humans will one day become subservient to machines.

Warwick's version of this claim requires, at the very least, that machines should be capable of intelligence. However, this is not an uncontroversial matter. The philosopher John Searle, for example, has famously argued, via his Chinese room thought experiment, that digital computational devices are not capable of meaningful thought, which, if he is right, leaves the issue of intelligence moot at the very least. So I start by asking Warwick what he takes intelligence to be, and whether he thinks that robots or machines are intelligent at the present time.

'Well,' he replies, 'some people, needless to say, want machines to exhibit human-like characteristics before they will consider them intelligent. The Turing test would be an example of this, as would Marvin Minsky's treatment of the issue in his book *The Society of Mind*. But, of course, the whole argument turns on what we mean when we use the word "intelligent". Is intelligence equivalent to knowledge, for example? In some ways, we seem to think that it is; take *University Challenge* or *Who Wants To Be A Millionaire?* – these are tests of knowledge. And we know that a computer, with its access to enormous knowledge bases, would win them hands down. So if intelligence has something to do with knowledge, if it has something to do with numerical ability, if it has something to do with communication, and so on, then machines *are* intelligent. But the important point is that they are intelligent in a *different way* from humans. Indeed, every creature is intelligent in slightly different ways. Bees understand the world by picking up water vapour signals. Humans can't. So any intelligence test which makes use of water vapour is meaningless for a human being, and human beings will be outperformed by bees. So I'm quite happy to say that machines are intelligent so long as it's clear that we're not saying the same kind of thing as when we say that humans are intelligent.'

Intelligence is, of course, a contested concept, so it is perhaps not objectionable that Warwick defines it in this kind of open-ended fashion. But there is nevertheless something about human intelligence which marks it out as being unique in the world; namely, that it allows for highly flexible, goal-oriented, non-context-dependent decision making. So is it likely that machines will become intelligent in the same kind of way as human beings?

'It is certainly likely that in the future machines will be able to accomplish any task which humans can,' Warwick replies. 'But that doesn't necessarily mean that they're intelligent in the same kind of way as humans. The fallback position for people who doubt that this is possible is to say that self-consciousness is something which machines will never have. This, of course, allows people to ignore all the data which suggests that machines are better than humans at accomplishing a great many tasks.

'My view is that machines probably won't ever be intelligent in exactly the same way as humans, unless the machine brain is made of the same stuff as the human brain, in which case, of course, it would for all intents and purposes be a human brain. But this whole line of thought is human-centric. Why start from human intelligence when thinking about machine intelligence?

'This is not to say that there aren't interesting philosophical issues to explore here. But even at the philosophical level, it just doesn't seem obvious either that machines will never have any kind of self-awareness, or that if they do not have self-awareness, they will necessarily remain less intelligent than human beings and that human beings will always be their masters.'

There are certainly reasons, even if they're not conclusive, to think that Warwick might be right that machine intelligence will one day match human intelligence. One thought here has to do with Moore's law, which, roughly speaking, describes how the power of computer processing increases over time. If Moore's law holds true over the next few decades – though there are doubts here to do with the limits of miniaturisation – then, in terms of raw processing ability, computer brains will be more powerful than human brains by about 2050. Of course, this does not necessarily equate to intelligence and consciousness; however, it is perhaps grounds for giving Warwick the benefit of the doubt when he is talking about as serious a risk as the possibility that machines will come to dominate human beings.

However, even granting machine intelligence, it does not seem obvious that this leads to the likelihood of human subservience. The author Isaac Asimov, for example, specified Three Laws of Robotics, designed to ensure that robot technology would be safe. These are:

1. A robot may not injure a human being, or, through inaction, allow a human being to come to harm;
2. A robot must obey the orders given it by human beings except where such orders would conflict with the First Law;
3. A robot must protect its own existence as long as such protection does not conflict with the First or Second Law.

(From *I, Robot*)

Why then, I ask Warwick, could we not simply implement these kinds of safeguards in any intelligent machines that we create?

'We could, but the chances are we won't,' he replies. 'But regardless, consider this scenario. You've got a financial machine. It's there to make a deal. If that deal means that it buys coffee from Kenya instead of Brazil, so be it. But, of course, that might mean that 200 people in Brazil lose their jobs, something it won't have been programmed to take into account. So a machine can inadvertently harm humans, even if you have something like Asimov's laws in place. And then think about military machines. It is absolutely ludicrous to think of putting Asimov's laws into a weapons system, since that would defeat their whole purpose. Any weapons system has an inverse Asimov law, *thou shalt kill.*

'If we give over decision making to *learning* machines, then we're in trouble. As soon as we allow machines to learn, to make decisions which they haven't been programmed specifically to make, then you can't be sure what they are going to do in any particular circumstance. Of course, this is only dangerous if we give machines the power to do things, like dropping bombs, and so on, but the chances are that we will; we want machines to be able to do things. Moreover, the risks are multiplied as soon as machines are part of a wide area network, because then they will be receiving information which we will have no idea about, and it will often be arriving from other machines over which we have no control.'

'The whole thing,' Warwick continues, 'is like walking to the edge of a cliff on a foggy day. Each step you take, you think, no problem, let's take another step, but in the end, you go over the edge.'

Warwick thinks that human beings should respond to this threat by radically embracing machine technology.

'It's the cyborg idea, the idea that humans should aim to enhance themselves by means of integral technology,' he tells me. 'As cyborgs we can give ourselves extra senses, the ability to think in different ways, the ability to communicate by means of thought alone, and so on. Hopefully in this way, we can secure the means to stay in control of the development of machine intelligence.'

That Warwick is serious about this idea is evidenced by two implant experiments which he has had performed on himself. In the first, in 1998, he had a silicon chip transponder implanted in his arm, by means of which he was able to interact with a computer – and thereby, lights, doors and heaters – situated at the Cybernetics Department at Reading. In the second experiment, conducted in 2002, a micro electrode array was surgically implanted into the median nerve fibres of his left arm. Using the array, Warwick was able to control an electric wheelchair, control and receive feedback from an artificial hand, and transmit signals to his wife, Irena, via a single electrode implanted in her median nerve.

Although Warwick thinks that these kinds of experiments are the first step on the journey towards a cyborg future, he admits that it isn't possible to see this future clearly. He talks about the possibility of things like 'communication via thought' and 'an enhancement of the senses', but concedes that it is very difficult to imagine what these things would be like. However, he does claim to have experienced first-hand an extra sense of sorts via one of the experiments which his team conducted with the array implant. The experiment involved a baseball cap with an ultrasonic transmitter and receiver. The transmitter sent out signals which were reflected back to the receiver whenever there was anything close by. This then resulted in a current being pushed down to Warwick's median nerve as a series of pulses. The closer the object, the quicker the pulses.

'It was just like having an extra sense,' he tells me. 'I was able to move around without bumping into things, because I just knew that they were there. The information was coming in as electrical impulses, but the sense that my brain made of it was just that there were objects nearby. It was an exhilarating experience.'

The problem with this kind of report is that it must be very difficult to discern whether what is being experienced is a pre-reflective awareness of the proximity of objects, or whether it is instead a cognitively mediated awareness that faster pulses indicate that objects are nearby.

'Yes, this is very difficult,' Warwick admits. 'It certainly seemed like immediate experience. But then we had been stimulating my nervous system in different ways, so I was tuned into the signals. Also, because the ultrasonic sensors were positioned on a baseball cap, and because I was blindfolded, it was like having replacement eyes, so no great leaps of the imagination were required. And, as you say, I did know what the pulses meant. But having said all this, what it felt like was just that there were things close by. So much so that when Iain, one of the researchers, came towards me with a board, it was just as if something was coming at me with great speed, and I couldn't see it.'

How then has this work been received? Has anybody else made this kind of deflationary criticism – that all that's happening with the extra sense is that there are impulses going into the nervous system, and because there is knowledge of what's going on, it is easy to work out that when the pulse goes quickly there is something close by, and when it goes slowly there isn't?

'I must admit that I've been a bit disappointed with the way that the work has been received,' Warwick replies. 'I thought that people might be more interested in the experiments with the extra senses. I would have been happy with the kind of criticism that you're talking about. But the response is either one of excitement or complete dismissal. I would have liked to have had a technical discussion about what was going on.'

'But I do realise,' he continues, 'that the lack of interest is partly a function of how I conduct my work. Nobody else is doing anything quite like this, and because it is a bit off the wall, it can be categorised as not being serious, or as not being science. I think it is difficult for people to respond with anything more than a knee-jerk reaction. But just because I crack jokes, and wear a baseball cap, and just because I like to try things out, it doesn't mean that what we're doing is not interesting.'

One of the more general criticisms which people make of Warwick's work is that it is over-hyped. For example, in a *Wired* article, technology journalist Dave Green complained that Warwick's predictions that robots will take over the world are combined with ineffectual demonstrations of their power. The example which he cites is the robot which Warwick's team designed to run a half-marathon.

'That's an interesting story,' laughs Warwick. 'The half-marathon robot, Rogerr – Reading's Only Genuine Endurance Running Robot – was entered first of all in the Reading half-marathon. Despite our working as hard as possible on it, when it came to the race, the robot wasn't ready. That was a shame, because there was a group of reporters ready to write about it. When Rogerr finally got to race six months later, at Bracknell, the race attracted twice as many people as usual because they wanted to run with the robot. It was a beautiful, sunny day, which we weren't expecting. All our testing had been done in grey, overcast conditions. On the Friday before the race, Rogerr wasn't working brilliantly, but we were confident it would get around. But it didn't even get a mile. The light from the Sun confused it, and it veered off and hit a curb!'

'But it was a good learning experience,' insists Warwick. 'It is a very hard thing to get to work. The problems are technical, rather than specifically to do with robotics. But unless you try to put one of these robots together, it is difficult to appreciate just how difficult it is.'

I put it to Warwick that in the case of Rogerr, the criticism seems a little unfair, since a lot of science progresses by trial and error.

'Exactly right,' he concurs. 'When I connected myself to the robot hand, it didn't work first time. But this is what research is about. When I first tried to switch a light on using my nervous system, the light didn't come on. It came on the second time we tried it. And the reports said: "He tried to switch a light on, and it didn't come on first time."

And I'm thinking "Hold on a minute, this has never been done before, and you're saying it didn't happen first time." '

However, notwithstanding the trial and error nature of much scientific endeavour, it is perhaps a fair criticism that some of Warwick's public pronouncements at least *seem* over-hyped. I mention the example of his claim that it was possible that other people might be able to experience his emotions by means of the micro-array implant. It just isn't at all clear how this could have worked.

'OK,' he replies, 'dealing with the specific question about whether or not it would be possible to experience another person's emotions, I had no real idea what would happen. When I wrote the article for *Wired* where I raised this as a possibility, I simply said that we would try.'

But surely there are good reasons for thinking it *wouldn't* happen. If our feelings and emotions are inextricably linked with our bio-chemical makeup, then it isn't clear how we can tap into them using a micro-array implant. Also, there is a more general problem here. Part of the promise of a cybernetic future is that it will be possible for people to communicate their feelings to each other directly. But if feelings and emotions are not simply electrical, doesn't this then become much more difficult? Although it might be possible to record the electrical outputs of emotional states, there doesn't seem to be a good reason for thinking that transmitting them back into a person will result in the same experience of emotion.

'Well, we just don't know,' Warwick responds to my objection. 'We have no idea. Probably you're right. People have different brains and nervous systems, and they change all the time. But that's not a reason not to try. If you're feeling sad, and we pick up the electrical signals associated with your sadness, then even though your sadness is electro-chemical, it is still an open question as to whether I will experience sadness, if we then blast the signals back into my brain. We don't know what will happen. And the only way to find out is to try.'

'Dealing with the more general criticism that I over-hype things,' he continues, 'I can see that there might be some truth in it. One reason is that I'm in the media a lot. Maybe I shouldn't say "yes" so often, but what I do is quite visual, and I get asked if I can do things a lot. The other thing is that we're in competition with American teams. If we're going to compete, then we have to market ourselves. People asked why we went over to the States to transmit signals over the Internet, when we could have done it in the lab. The answer is that we were competing with a US team who had got a monkey to move a robot arm which was in a different building. If we'd stayed in our lab, then people would have complained that what we had done wasn't as impressive as what the American team managed.'

This response is interesting, because it raises the issue of funding, and its relationship to the kind of work which Warwick is pursuing.

'In an applied science department you're never going to get large amounts of money through your students,' he explains. 'It might be slightly different with a subject such as sociology where you don't have huge outgoings, but we need the latest equipment, we need money for student projects, and so on. So it is necessary to bring in outside money; there just isn't enough from the government.

'When I first came to Reading, I decided that there were two avenues we could pursue. One was linking up with local industry. We were very lucky that in the Thames Valley area there were a lot of companies doing pretty well. The other avenue was to increase student numbers. That wasn't easy because cybernetics just doesn't have the history or profile of something like physics. People don't study it at school. The only way we could get students was to market cybernetics directly. And to market the subject we needed to get people aware that it existed. And cute robots seemed to be a good way to do this, so we built them.'

However, whilst Warwick is happy to admit that publicity is important for attracting funding, he is adamant that funding consider-ations do not drive his cyborg research. 'I have my own itinerary, and I follow it,' he insists. 'The funding that we have attracted has in no way changed or modified the cyborg project as originally conceived. Some-times I've worn a shirt or a hat with a sponsor's name on it, and I've gone on to the television, and that's fine. I give presentations to com-pany executives from time to time, and sometimes to a sponsor's cli-ents. And again I don't have a problem with this. But the overall direction of the research is not affected by the demands of the sponsors.'

What makes this response entirely plausible is that Warwick is quite clearly driven by a deep curiosity about what it would be like to be a cyborg. He admits to being an old-style, give-it-a-try experimenter. 'When we were injecting current onto my nervous system, we had no idea what was going to happen. It was just tremendously exciting. When I was a kid, I was excited by things like Scott going to the Antarctic, and JFK saying that people were going to go to the Moon, and there's still this aspect to me. There's so much unknown about the world. And science is potentially tremendously exciting. I don't really understand how some scientists are content with a kind of plodding routine. Are they really excited when they get up in the morning? I have a challenge to myself. If any morning I don't run up the stairs to my office because I'm no longer desperately keen to get on with things, then that's the time to quit. And every morning I bound up my stairs!'

There is an interesting ethical question raised by Warwick's evi-dent enthusiasm for his cybernetics research. In his book *I, Cyborg*, he speculates that humans who choose not to go down the cyborg path will become a kind of subspecies. Isn't this a good reason – even if it isn't a conclusive reason – for approaching the kind of research he's doing with caution? Perhaps for not bounding up the stairs quite so quickly?

'It's about allowing people a choice,' he answers. 'If people want to stay as they are, then fine. But I don't. I can see that some people may want to stop my kind of research, but I think I should be given the option to take the path that I choose.'

'But, having said that,' Warwick continues, 'one of the reasons that I put it in these stark terms is to stir up debate. Last year, for example, I was contacted by lots of parents who wanted to know whether I could tag their children, so that they could trace them if they were abducted. I said that I would try to put something together. A news agency then contacted me, and we told them what we were up to. There were then headlines that there was a particular 11-year-old girl who was going to get an implant of some kind. This meant that there was a case to discuss, and the ethical issues could be debated. The response from various children's charities was very negative. They said that it wasn't a good idea, that it would not do the children any good, that they weren't being given a choice, which made it assault, and so on. I was in the middle of this, but I shared their concerns. There was no way that I wanted to go ahead without going public, precisely because I wanted the ethical debate out in the open. I listened to what they had to say, and I decided not to go ahead. So with all these things there are ethical debates to be had.'

Back in the year 2000, on the BBC's *Start the Week* radio show, Jeremy Paxman told Warwick that he was either a visionary genius, or a publicity-crazed lunatic, and that he wasn't sure which. Probably we won't know the full answer to this question until the twenty-first century has played itself out. However, there is no doubt that Warwick is prepared to countenance what to most people would be unthinkable.

I ask him what ambitions he has left for his cyborg research. This is what he tells me.

'I would like to link my brain up to a computer. We're probably talking about a ten-to twelve-year time frame. I'm not going to retire until I've had a go at it. We're looking at putting some kind of port into the motor neuron area. And we're building up to that, deciding where in the brain the implant should go, how many implants, what kind of implants, and so on. That's my long-term goal.'

In conversation with Robin Murray

Robin Murray, Professor of Psychiatry and Head of the Division of Psychological Medicine at the Institute of Psychiatry, in the UK, was a teenager when he first decided that he wanted to be a psychiatrist. But his decision was not greeted with unqualified enthusiasm.

'I remember at school,' he tells me, 'that our housemaster and his wife asked us what we wanted to do, and various boys said they wanted to be lawyers, businessmen, dentists, and so on, and each was greeted with great encouragement. Then I said that I wanted to be a psychiatrist. The housemaster didn't respond, but his wife said, "But Robin, your parents are such normal people." I guess her reaction was an indication of the stigma attached to the profession. It was not only the mentally ill who were frightening, but also those who might want to get involved in their treatment.'

The stigma remains to this day. Partly, as Murray indicates, it has to do with the fear which people have of mental illness and the mentally ill. But it is also a function of the perception that psychiatry is a discipline not in full control of its subject matter. This perception is encouraged by tabloid headlines which scream about the supposed shortcomings of psychiatrists whose patients have committed acts of violence whilst in their care. And it is also fuelled by a suspicion that psychiatry is in the business of inventing spurious mental disorders.

The *Diagnostic and Statistical Manual IV* (DSM-IV), for example, widely used by psychiatrists, especially in the United States, contains nearly 400 mental disorders, including: oppositional defiant disorder, a childhood condition characterised by, amongst other things, loss of temper and arguing with adults; and malingering, which involves the intentional production of false or exaggerated physical or psychological problems.

The practice of classifying patterns of symptoms into illness categories was begun by German psychiatrist Emil Kraepelin, at the end of the nineteenth century. How then have we progressed over the last hundred years to a situation where we now have close to 400 separate disorders?

'Well, I'm not sure that is progress,' responds Murray. 'It has changed, certainly. European psychiatry has stuck reasonably closely to the Kraepelinian model. However, in the United States, largely because of the influence of the psychoanalysts, the scheme broadened out. At the beginning of the twentieth century, the focus of psychiatry was Germany. But with the rise of Nazism, many of the liberal, Jewish psychiatrists fled. Those who were most influenced by science came mainly to Britain, but those influenced by Freud and psychoanalysis went to the States, where they made a lot of money. And one of the ways that they made money was by expanding the boundaries of psychiatric illnesses. By the 1960s, three times as many people were being diagnosed as schizophrenic in the States as in Europe. Fortunately, though, such disparities are a thing of the past as the use of more precise definitions has meant that rates are now internationally comparable.'

The fact that it was possible to find such a great difference in the rates of diagnosis of schizophrenia points to one of the criticisms which people make of the illness categories of classificatory schemas such as the DSM; namely, that they are thoroughly culturally embedded.

Is there anything to this criticism that mental disorders are, at least in part, social constructs?

'Yes, it is valid,' says Murray. 'But it is important to understand why the DSM was developed. It was put together because whilst there was a very broad concept of mental illness in the States, it was not one where categories were clearly defined. There was, for example, a Manhattan survey which claimed that about 90 per cent of the population had a psychiatric disorder. What happened was that in the 1960s and 1970s, the anti-psychiatrists said "Look, psychiatrists can't define these disorders" and "It's not the patients who are mad, it's society which is mad." The reaction to this was a swing back to the Kraepelinian view that there were specific disease categories, and a lot of emphasis was put into making the diagnosis of these more reliable. The development of the DSM was part of this move away from psychoanalysis towards medicine.

'You might, therefore, think that the DSM categories were developed by scientists on the basis of research advances, and this is partly true. But they were also put together by powerful men in smoke-filled rooms, engaging in trade-offs with each other: "I will support your criteria for anorexia nervosa, if you support mine for depression".

'All this was also pushed forward by the insurance system, because in the States you cannot get reimbursed for treatment for general social distress. Low self-esteem, for example, is not a reimbursable condition. But if it is classified as minor depression, then reimbursement is possible. So there was a political and economic dimension to this process as well.'

Does this then mean that many of these disorders have no underlying pathology, or that if one did a factor analysis of multiple sets of symptoms, it would not be possible to distinguish between different disorders?

'Yes, I think that's right for some of the categories in the DSM,' responds Murray. 'The classificatory system is a mixture of entities, dimensions and political imperatives. In addition to the influence of the insurance companies in the States, we now have a lot of pressure from the drug companies. These companies want to expand what is regarded as a disease to enlarge their market. And since psychiatric disorders have ill-defined boundaries, they are an excellent playground for the drug companies to invent new diseases. So psychiatrists some-times joke that there will soon be a post-vacation dysphoria disorder – post-holiday blues – and there will be a drug for it.'

Although the DSM introduced a myriad of new sub-disorders into the psychiatric firmament, Murray is optimistic that a better evidence base will provide a more rational approach to classifying disorders. He points to the new knowledge of brain function in health and disease which is flooding into the clinic from neuroscience, genetics and brain imaging. And he believes that the challenge of studying the mind is now attracting the best young researchers and cites advances in many areas including his own field of schizophrenia, or *dementia praecox* as it was in its Kraepelinian formulation.

Schizophrenia is probably the best known, and most notorious, of the mental illnesses. How, I ask, does a clinician go about diagnosing schizophrenia? What kinds of symptoms do they look for?

'The classic symptoms are hallucinations, particularly auditory hallucinations – hearing voices when nobody is there,' he replies. 'This tends to start in a simple way, perhaps with a name being called out, but the more ill a person becomes the more systematised the voices become, and the end result is often a series of voices talking about the sufferer. One may be saying: "He's a shit, he should kill himself." Another will be saying: "It's not his fault, they are out to get him." Obviously, when a person hears a voice, they will think: "What's the cause of this?" or "Where's it coming from?" And because there is no obvious answer, sufferers begin to make up delusional explanations as to the source of the voices. So the beginning of schizophrenia is often a simple auditory hallucination, but delusions then emerge – for example, that the CIA are responsible – in order to explain the hallucinations.'

Most people are aware that the mentally ill will often hear voices when there is nobody speaking. But what is less well known is that these voices are not 'in the head', as such, but rather seem to be coming from the outside world.

'When people are acutely disturbed, they hear these voices as clear as a bell, as if it were someone talking to them,' Murray explains. 'Indeed, often it will seem as if the voices are coming at them through a loud-speaker. As they begin to recover, people may say that the voices aren't so loud, or that they seem to be further away, and as they get better, they will often say that the voices seem to be in their head, that they don't seem to be from the outside. And then when they're really improving, they say that sometimes they don't know whether it is a voice or a thought. They begin to integrate the voices back into their thoughts.

'Voices are a misinterpretation of inner speech. When a non-schizophrenic person thinks of his favourite poem, he may well say it silently in his mind, but he knows that he is producing the words of the poem. A person with schizophrenia produces the words in their mind, but misidentifies them as coming from an external source.'

In this connection, Murray mentions a study by the British psychologist Chris Frith, who was interested in whether people prone to misinterpreting inner speech would also have problems recognising their own external speech. Frith got subjects to speak into a microphone, and then fed a distorted version of their own voice back to them. When people without psychosis were asked what was going on, they knew that their voice had been distorted. However, people who had suffered hallucinations began to think that something strange was happening quite soon after their voices were first altered.

Part of what is interesting about this kind of research is it highlights the role of delusions in explaining the meaning of what seem to people with schizophrenia to be puzzling experiences. Murray explains, as follows: 'Frith gives an example of one patient who said that his distorted voice was the voice of his brother speaking. He said that his brother didn't want him to know that it was him speaking, and therefore disguised his voice. Frith then explained to this patient that he – the patient – was speaking into a microphone, and that his voice was being distorted by a special effects box and fed back to him. He asked the patient whether he understood what was happening. "Yes, Doctor," he replied, "I understand everything. I speak in here. My voice is distorted by a box. And I'm hearing a distorted version of my own voice. It's not somebody else or my brother speaking. The only thing I don't understand is how does the box know what my brother sounds like?"

'So,' Murray continues, 'people have these strange experiences, they search for meaning in them, and the delusions arise in the context of their belief system to explain their experiences, and probably to reduce the anxiety associated with having such experiences.'

One interesting thing about this process is that people don't jump to what would seem to be the obvious conclusion when they have these hallucinations: namely, that they've got some kind of mental illness. After all, it probably is the case that most people know that mental illness is associated with hearing voices. Isn't it a bit odd that they can't draw upon this knowledge?

'Well, I can think of a chap who was in the middle of doing a BSc in psychology when he started hallucinating,' Murray responds. 'He heard voices speaking to him. He thought that other people were out to harm him. When he recovered, after getting treatment, he said that it was amazing that there he was, a psychologist, and he knew all about the symptoms of schizophrenia, and when it happened to him it never crossed his mind that he might be suffering from the illness. He *knew* they were out to get him. He *knew* that these voices were real. Presumably it is something about the nature and intensity of the experiences which makes them so hard to resist. But we also now know from brain imaging that when sufferers are hallucinating they are activating not only the area in the brain which produces inner speech, which is called Broca's area, but also the auditory cortex in the temporal lobe which is normally involved in processing external speech. The sufferer is tricked into believing that the words are coming from an external source because they are being processed by the same neural system that we use when listening to external speech.'

The predominant view now about schizophrenia is that its symptoms – auditory hallucinations and the like – are produced by a dysfunction of the brain. The dopamine hypothesis suggests that the problem is mediated by an excess of dopamine transmission, particularly in the mesolimbic system of the brain. There is a lot of evidence to support this hypothesis. For example, all anti-psychotic drugs work by blocking dopamine, and all drugs that increase dopamine can cause psychosis. Furthermore, brain imaging shows that if individuals are given amphetamine, non-schizophrenics will produce only a small amount of dopamine as a result, whereas schizophrenics will produce a flood of dopamine – and the bigger the flood, then the more psychotic the individual.

However, the fact that schizophrenia is associated with a particular kind of brain dysfunction tells us little about the origins of the illness. For example, it leaves open the question as to whether schizophrenia is a degenerative disorder or a developmental disorder. What, I ask Murray, is his view about this?

'The orthodox view used to be that schizophrenia is a degenerative disorder – that sufferers start off with normal brains, and something happens to the brain in their teens or twenties, and then they gradually deteriorate. When the brain scan era started, subtle differences between the brains of schizophrenics and non-schizophrenics were found. So it was assumed that the disease was degenerative.'

'However,' continues Murray, 'in the mid-1980s, my colleagues and I, in Europe, and Daniel Weinberger, in the States, began to think that these differences were not a sign of deterioration, but rather an indication of a developmental problem. The brains of people with schizophrenia had somehow developed in a subtly deviant way. In some way, the wiring had developed imperfectly, and they were therefore more vulnerable to hallucinations and delusions. And really from the mid-1980s to the present day, this has been the dominant view.

'However, as often is the case, the world has over-swung. I never said, or at least I hope I didn't, that schizophrenia is *just* a developmental disorder. I think there is a developmental component to it. One of the exciting things about working on schizophrenia is that it is at the crossroads of neuroscience and social science. So whereas some believe that schizophrenia is simply a brain disease, my view is that many people with schizophrenia have developmental difficulties which make them less able to cope with events which they then encounter in their lives.'

How then does all this fit together with the dopamine hypothesis? Presumably there isn't a conflict here?

'That's right, there isn't,' confirms Murray. 'The developmental view would be that the affected individual either inherits a dopamine system which is more unstable than normal or else things happen to his or her brain during development which renders the dopamine system more unstable. For example, we know that individuals who are born small, or who have suffered hypoxic damage – loss of oxygen – either when they're *in utero*, or at the time of birth, are relatively prone to psychosis. As the cortex of the baby is developing in the latter part of gestation, a whole series of complicated neuronal patterns is set up. If some insult happens at this time, it can interfere with the development of the cortex, which may result in neural systems which are not quite perfect – and consequently, brains which are probably not quite so adept at handling situations of stress. So there isn't any contradiction between the dopamine hypothesis and the developmental hypothesis. The developmental hypothesis suggests that either for genetic reasons or because of early environmental insult, neural systems develop in such a way that stresses to the dopamine system are likely to be much more catastrophic in that individual.'

Murray mentions here that genetic factors are involved in the development of schizophrenia. The evidence for this is long established. For example, the first twin studies showing an inherited component to schizophrenia were completed more than seventy years ago. But what about the advances in molecular genetics of the last twenty years, have they led to a greater understanding of genetic risk factors?

'I think it is fair to say that until very recently the molecular genetics of schizophrenia was a disaster area,' Murray laughs. 'I used to say that there was no area of human endeavour, apart from the search for alien spacecraft, which had produced more false positive results than the molecular genetics of schizophrenia! Whenever some research group claimed that they had found a gene for schizophrenia, another group immediately contradicted them.

'However,' he continues, 'this has been useful in a sense, because twenty years ago many experts thought that there might be a big gene for schizophrenia; that is, that you might inherit a gene for schizophrenia which would render you almost certain to develop the illness. This is the "doomed from the womb" hypothesis. But we now know that there are no genes which condemn people like this. Instead, as in coronary artery disease or diabetes, it is likely that there are a number of susceptibility genes which interact with each other and with environmental effects.

'And in the last two years, there have been dramatic developments as two genes have been identified which do seem to increase the risk of schizophrenia. An Icelandic group reported a gene called neuregulin on chromosome 8, which has now been replicated by four other groups including ourselves. And an American group claimed to have found a gene called dysbindin on chromosome 6, which, so far, three other groups have replicated. However, these are genes of very small effect; if you carry one of these genes your chances of developing schizophrenia may be doubled. They do not result inevitably in schizophrenia, but if you carry these genes, plus other susceptibility genes, and suffer a number of environmental effects, then you may develop the disease.'

But don't the twin studies suggest rather a large genetic effect?

'If you take identical twins, and one has schizophrenia, then the second twin will be psychotic in about 50 per cent of cases,' Murray confirms. 'Furthermore, a proportion of the co-twins who escape schizophrenia will not be entirely free of mental illness. So there is undoubtedly a big genetic component. But it is a bit like cardiovascular disease. If you find an identical twin who has had a heart attack, then about 40 per cent of the co-twins will also develop heart disease. But this is not to say that developing heart disease is not influenced by smoking, obesity, taking no exercise, high blood pressure, and so on. It's the same with schizophrenia. Identical twins inherit the same genes, which makes them vulnerable, but environmental factors are important too.'

The importance of social factors in understanding schizophrenia is something which Murray is very keen to stress. He talks at some length about how the research which he and his colleagues are doing with African–Caribbean patients is shedding light on some of the interesting ways in which these factors might be involved.

'We know that people of African–Caribbean origin are about six times more likely to be diagnosed with schizophrenia in the UK than the white population,' he explains. 'The question is why? Is it genetic? Is it because of discrimination? The first thing we considered was that it might be misdiagnosis. Perhaps white psychiatrists are unfamiliar with the cultural norms of their black patients, and might, for example, misconstrue a belief in witchcraft in a person brought up in the Caribbean as a delusion. However, we found the same increased rates in people of African–Caribbean descent who are born in Birmingham or Brixton. They're still six times more likely to develop psychosis. And also when our patients were re-diagnosed by a Jamaican psychiatrist, he came up with roughly the same proportion as having schizophrenia, so it isn't simply a matter of psychiatric bias.

'We then wondered what the rates of schizophrenia were in the West Indies. So colleagues carried out studies in Jamaica, Trinidad and Barbados, and they found that the rates were just average there. So not only are the rates of schizophrenia amongst people of African–Caribbean origin living in the UK six times higher than the white population in the UK, they're also six times higher than the black population in the West Indies.

'We then thought that we would look at the families of these individuals. When we look at the parents of white people with schizophrenia, we find that some 5 per cent of them also have schizophrenia. When we looked at the parents of African–Caribbeans with schizophrenia, it was the same, so it wasn't as if there was a greater genetic predisposition.

But when we looked at the brothers and sisters of African–Caribbean people with schizophrenia, some 6 per cent of the brothers and sisters living in Jamaica had the illness, compared to about 26 per cent of brothers and sisters living in the UK.

'So these are individuals who have some vulnerability, but if they were still in the environment of the West Indies, then they would be far less likely to develop schizophrenia than is the case if they are here in the UK. There is something about the environment here which increases the risk of schizophrenia. What we have found is that there is a lot of adversity in the lives of these individuals at the time at which they become psychotic. There is an important point here. These individuals have the symptoms we call schizophrenia in the white population, but the fact that they have these same symptoms does not necessarily mean that they have the same biological causes. So it is possible that you can get the schizophrenia syndrome for a mixture of reasons.'

Murray is alluding here to some of the complexities of the relationships between symptoms, causes and illness. Indeed, these kinds of complexities have led some people to question the usefulness of thinking that schizophrenia constitutes a single clinical entity. Perhaps most notably, anti-psychiatrists such as Thomas Szsas, in the 1960s and 1970s, suggested that schizophrenia in particular and mental illness in general simply do not exist. Roughly speaking, their argument was that people are labelled as suffering from a disorder such as schizophrenia in order to 'control' and 'neutralise' behaviour which is deemed to be socially unacceptable. Is there still an argument to be had here or does the evidence from inheritance studies, the effectiveness of anti-psychotic drugs, and so on, point overwhelmingly to the reality of mental illness?

'Well, first, I don't think anti-psychiatry exists as a significant force any longer,' responds Murray. 'It was an important phenomenon which

I think was a reaction to the bad psychiatry that was going on. It's interesting that today some of the most radical critics of the concept of schizophrenia are amongst the most able academic psychiatrists. Indeed, I would be surprised if in twenty years' time, we haven't unpacked the umbrella term schizophrenia into its differing components. So in many ways the criticisms of the anti-psychiatrists have been taken into the mainstream view.

'Also, a lot of the anti-psychiatry comment that is still around comes from the survivors of bad psychiatry, and I think it is entirely justified, because our psychiatric services are very bad. They have been so starved of resources that most in-patient psychiatric units in London, for example, are horrendous places. They'd be the last place you would want to go if you were very distressed and you wanted some peace and quiet. So there are justifiable criticisms of the quality of the care which people get. Society doesn't care enough for the mentally ill to ensure that they get treated in reasonable surroundings.'

But, nevertheless, surely there are specific claims which anti-psychiatrists make which don't stand up against the evidence? For example, what about the idea that the reason that schizophrenics show brain abnormalities is because they have been treated with powerful anti-psychotic drugs?

'Well, again, it's not entirely clear,' Murray cautions. 'First of all, there is a question about whether the differences we see in the brains of schizophrenics are abnormalities or deviations. You can't look at a person's brain and determine whether or not they have schizophrenia. But having said that, the brains of schizophrenics *are* statistically different from the brains of non-schizophrenics. And some of the deviant characteristics *are* there before the onset of schizophrenia. Indeed, some of these characteristics are present in the relatives of schizophrenics. But these characteristics are *risk* factors for the illness, they are not necessarily part of the illness itself, rather like the way in which having a high cholesterol is a risk factor for heart disease.

'We also know that some of the older anti-psychotic drugs are associated with brain changes. Many of them can cause movement problems, and these have been linked to an increase in the volume of certain of the brain structures which are involved in motor control. So it is true that some of the old drugs, given in high levels, over a long period can induce brain changes.'

But in terms of how the evidence stacks up overall is it not the case that it suggests that there are brain abnormalities present in schizophrenics prior to any intervention by health professionals?

'Yes, but the picture *is* complex,' insists Murray. 'At one point you had the anti-psychiatrists saying that all of schizophrenia was socially determined, and that the idea that there was a biological substrate to psychosis is ridiculous. And equally, you had the crude biological psychiatrists saying that they could see big ventricles in the brains of people with schizophrenia, and that these individuals must have a degenerative brain disease. Thirty years on, we now know that both these points of view were inadequate, though there was some truth in both positions.'

Although Murray thinks that anti-psychiatry was an important corrective to the excesses of an overly biological approach to schizophrenia, he does not have any sympathy with the more radical claims of the anti-psychiatrists. Particularly, he considers that the views of R. D. Laing that the parents of children with schizophrenia were in some way responsible for the illness were deeply damaging.

'It was bad enough,' he tells me, 'that you had one of your children develop psychosis, but then to be accused by a psychiatrist or social worker that you had contributed to the disease was just horrendous. Twenty years ago, one would see a lot of parents whose lives had been destroyed not so much by the illness of their child, as by the accusation that they had caused the illness. So I think the anti-psychiatry movement caused a great deal of harm, without much evidence to back up its theories. But there was nevertheless a kernel of truth in what they said. And to some extent we have now incorporated their criticisms into mainstream psychiatry.'

There's an interesting point about labelling here. Anti-psychiatry resulted in many parents being labelled as the cause of their children's schizophrenia. But is it possible, with the explosion of diagnostic categories, especially in the States, that things might now be working the other way around? Attention deficit disorder (ADD), for example, might be something that parents would want their child to be diagnosed as having, if the alternative was that their child's behaviour was seen to be a product of their home life, and so on.

'I'm sure that's true,' Murray agrees. 'The expansion of hyperactivity and ADD has been driven by society, parents, and by the drug companies. Of course, it has been exploited by some psychiatrists, and this is particularly the case where you're talking about the private practice system. There is no incentive for the average American child psychiatrist to say to parents, your child is normal, you should go away, and try to alter the social circumstances you're bringing your child up in. There is a great deal of economic pressure for the psychiatrist to say, I will solve this biologically. So ADD is diagnosed much more commonly in the US than the UK. However, the diagnosis is becoming more common in the UK. Presumably the gene pool hasn't changed, the brains of growing children haven't changed, but in some way the expectations of parents and society have changed.'

What about treatments for a disease such as schizophrenia? If a person was diagnosed with the illness sixty years ago, then the prognosis was pretty bleak. Almost inevitably patients were confined to mental institutions, in wards which were referred to then as 'snake pits'. But presumably this has changed with the emergence of anti-psychotic drugs?

'Yes, it has,' confirms Murray. 'The sad thing about asylums is that they were set up with the best of intentions. In the Victorian era of moral therapy, the idea was that people should be taken out of the inner cities where they often became ill, and that they should go to live in the country where they'd get fresh air and proper food, and be well looked after, before going back to their normal life. The difficulty was that a proportion of patients did not recover, and therefore asylums gradually filled up with more and more people, the staff became more and more overworked, and eventually they simply became institutions of containment. So, before the Second World War, a very significant proportion of the people who went into these asylums would stay there.

'But with the development of anti-psychotics, the emergence of community care, and the arrival of rational psychological treatments, things have improved a lot. Many people now do remarkably well. I have patients who are successful professionals, who get on with their lives without any of their colleagues knowing that they have a diagnosis of schizophrenia. Provided they get regular support, and psychological help at times of crisis, they do very well.'

Although it is true that the emergence of anti-psychotic drugs and new psychological treatments, particularly cognitive therapies, has helped patients with mental disorders, there are a number of worries which the public have about the way that the psychiatric profession operates. One specific worry, in the UK at least, is about the community care programme, the practice of managing illnesses like schizophrenia in the local community rather than in institutions. What, I ask Murray, is his view about the community care programme? Has it been a success?

'Since the 1950s,' he replies, 'most Western countries have moved away from a hospital-based system, and there are some countries, like Holland, where this has worked extremely well, and there are other countries, like the UK, where it has worked rather badly. We were unfortunate that the big push towards community psychiatry occurred at the time of the Thatcher cuts. The theory was that the money we'd save by shutting the asylums would go into developing community placements and community facilities. Unfortunately, because the health service was in crisis, the money was diverted into propping up intensive care units, and that kind of thing. Therefore, there were never sufficient resources devoted to ensuring adequate care of people in the community. Then there was a reaction in society at large, fuelled by the media, which saw people becoming very suspicious of patients in the community, and psychiatrists becoming increasingly anxious about the prospect that any of their patients might be violent and that they might be held responsible.

'This is a very big difficulty. Most psychiatrists have a friend or a colleague who has been on the front page of one of the tabloids, because a patient of theirs has been involved in a violent incident. You can be a wonderful psychiatrist, and practise brilliantly for thirty years, but if you have one patient who commits a violent act, and this is taken up by the tabloids, then you're in trouble. We're in a very difficult situation at present. We have a government which is increasingly authoritarian in its dealings with the mentally ill, a government which, out of the blue, concocted a new psychiatric diagnosis – "dangerously severe personality disorder". This was driven by the media and politicians. There is no such condition as "dangerously severe personality disorder", but units are being set up across the country to deal with it. We know that certain groups are more likely to commit violence. But the evidence is that in order to prevent one violent incident you might have to permanently lock up nine individuals. The current government policy is going to result in the psychiatric services being used as a branch of the criminal system. Perhaps the only good thing about their policies at the moment is that they have done a great deal to heal any rifts between psychiatrists, patients and carers. Just about everyone who has anything to do with the care of the mentally ill despises the proposed new Mental Health Act.'

7   Microbiology, Viruses and Their Threats

In conversation with Dorothy Crawford

In the mid-1980s, microbiology hit the headlines in a big way with the rise of the AIDS epidemic. Since then, microbiological illnesses have rarely been out of the news: salmonella, legionnaires' disease, Ebola fever, necrotising fasciitis, anthrax, smallpox, foot-and-mouth disease, West Nile virus, MRSA and SARS have all been front page stories in the last twenty years. However, despite this coverage, there is still much confusion amongst the general public about the nature of the microbes which cause these illnesses. Particularly, many people are unaware that viruses and bacteria are very different things.

'This is a bit of a hobby horse of mine,' says Dorothy Crawford, Robert Irvine Professor of Medical Microbiology at Edinburgh University. 'It is really very irritating that the media tend to label every nasty illness a virus. Viruses are actually very different from bacteria. In fact, about the only thing which they have in common, apart from the fact that they cause illness, is that they're very small. But even here, viruses are in fact much smaller than bacteria; for example, you can't see viruses without an electron microscope, whereas you can see bacteria with an ordinary light microscope. Indeed, until the electron microscope was invented, people weren't at all clear about what viruses were; they thought that they were probably just very small bacteria, but in fact they're very different.'

'Bacteria are single-celled organisms, which have the ability to exist independently of other organisms,' Crawford explains. 'They are able to metabolise and to reproduce on their own. Viruses, on the other hand, are *obligate parasites*; they are only able to reproduce within a host organism. They can't do anything until they get inside a living cell. Once they're there, they have mechanisms which enable them to take over the cell, and to use the cell's organelles in order to reproduce themselves. They have nucleic acid – DNA or RNA – so they have inherited characteristics; but apart from that there isn't really much to them.'

The fact that viruses do not produce their own energy and cannot reproduce without hijacking the cells of a living host raises the question as to whether they are alive or not. Does Crawford have a view about this?

'Well, it's a question which I sometimes ask my students,' she replies. 'It's a good debate, but I'm not sure how much it matters. Ultimately it hinges on how life is defined. If life is about reproduction, then viruses are alive; but if it is about the ability to metabolise, then they're not. I guess, if pushed, I'd come down on the side of viruses not being alive. They just don't seem to have lifelike characteristics.'

If viruses then are just sets of genetic instructions, how do they cause illnesses? Presumably it has something to do with the way that they hijack a host's cells?

'Yes, that's right,' Crawford confirms. 'A virus is just a piece of genetic material which will replicate itself if it can. If it can't, then it won't survive, so, in that sense, reproduction is what viruses are all about. If a virus gets into a cell, and it finds all the material it requires, reproduction occurs automatically via biochemical reactions. The consequence is a lot of new viruses. This process upsets the host cell, and will very often kill it. So, if a flu virus, for example, gets in via a person's nose, and infects their respiratory epithelium, then, if enough copies are produced, killing a lot of host cells, the infected person is going to feel pretty ill.

'However, not all viruses work this way,' Crawford says. 'Other viruses, for example, go into a cell, and then stimulate it so that the cell replicates itself. The virus requires the enzymes which are found in the cell when the cell is growing. This process, if it is not checked, can result in a cancer; there is a breakdown in the normal checks on cell division, so unless the immune system notices what is going on, you can get a tumour.'

Viruses, of course, can cause huge numbers of deaths. For example, the influenza pandemic in 1918 is estimated to have killed 40 million people. However, most viruses do not have this kind of impact. Is there something about what it takes for a virus to be successful which makes really high casualty figures relatively unlikely?

'The key thing is how the body responds to a virus,' Crawford answers. 'The 1918 flu virus was completely new, so nobody was immune to it. Generally speaking, the human body is not prepared for completely new infections, so you get a worse illness the first time a particular virus comes around. In evolutionary terms, those who are most susceptible to new viruses die out; therefore, they don't produce offspring, which means that populations become more resistant to specific viruses over time. This would likely be the case with the SARS virus. If it keeps turning up, then eventually people will have more resistance to it. But, of course, there is such a long time between generations of humans that it would be more than a century before any significant resistance emerges.'

What about an illness like Ebola fever; isn't it just too deadly to be a major threat?

'Probably that's true,' Crawford confirms. 'If you're infected with Ebola, then you're too ill to move anywhere, so whilst it's an explosive outbreak, it doesn't spread very far. Also, in the case of Ebola, its method of transmission goes against it. It is spread by close contact, whereas the viruses which result in pandemics are spread through the air. Having said that, though, something like HIV is very successful, because it is so insidious; it is possible to be infected and not to know it. It also exploits a sure-fire mechanism for going from one individual to another. So whilst it is slow spreading, compared say to flu, it is extremely difficult to eradicate.

'The other point to make is that part of the danger of viruses is that they can be unpredictable. For example, when the H5N1 avian flu first appeared in Hong Kong in 1997, it wasn't clear how it would develop. Although it killed quite a high proportion of the people it infected, it turned out in the end that it wasn't very infectious; it wasn't able to spread very quickly. The authorities in Hong Kong were quick off the mark, and by killing thousands of chickens they were able to prevent it from running out of control. It re-emerged in Hong Kong and the Far East in 2003. Again, it killed large numbers of chickens but not many humans. Thankfully it looks as if it isn't the kind of virus which is going to cause a pandemic. But this wasn't clear at first, and for a while it was quite frightening. Of course, the chances are that someday a new strain of flu will emerge which does have the potential to cause a pandemic, but we now have very sophisticated monitoring mechanisms in place, so just maybe we will be able to do something to stop it spreading.'

Given the dangers which influenza poses, and given the length of time that the virus, in one form or another, has been around, what has prevented us from developing a vaccine which is effective against all its strains?

'The flu virus changes the whole time, it's constantly mutating, and the changes which occur can be very large; that's gene swapping, rather than just single base pair swapping,' replies Crawford. 'This means that all you can really do is to second-guess which particular strain is going to appear twelve months down the line. The World Health Organization (WHO) produce a vaccine based on what they think will be the most prevalent strain of flu. But, of course, if a completely new strain turns up, then it takes time to create a new vaccine, by which point it may well have spread a long way. The policy we have at the moment is a sensible compromise: we vaccinate on the basis of what we think is likely to be around, and we target old people and people with chronic illnesses, because they're at most threat from the virus.'

The threat posed by viruses such as influenza has been exacerbated in recent times by the huge growth in international air travel. The outbreak of SARS which occurred at the beginning of 2003 illustrates this point nicely. The illness was largely confined to areas in East and Southeast Asia. However, there was an outbreak in Toronto, Canada, which it proved possible to trace to one woman – Sui-Chu Kwan – who had contracted the disease during a visit to Hong Kong.

'International travel does have a huge impact on the potential for a virus to spread,' Crawford confirms. 'But the speed with which the authorities responded to SARS was pretty impressive. Also, it is likely that the effects of globalisation will be somewhat mitigated by the monitoring processes which are now in place. But, in the end, a lot of this comes down to chance; sometimes you can just be unlucky.'

If there is an outbreak of a disease such as SARS, what kind of strategy is it best to pursue in order to keep it in check?

'Well, it partly depends on whether you have a vaccine or not,' Crawford replies. 'In Toronto, for example, if there had been a vaccine for SARS, then they could have vaccinated all the people who had

had some contact with the first victim. The idea is to ring-fence the infection in an attempt to stop it spreading. However, there is always some dispute about what kind of policy is most likely to be effective. For example, during the recent foot-and-mouth disease outbreak in the UK, there was a huge amount of debate about whether vaccination was appropriate; whether it would make the problem worse or better. After the outbreak was over, they looked at the issue using some very sophisticated mathematical models; and what the models suggested was that if they had ring-fenced the infected farms, and used targeted ring vaccination around them, then that probably would have cut the duration of the epidemic. But the problem is that if an epidemic springs up very quickly, it is only really possible to do this kind of modelling after the event.'

There has, of course, been some success in combating viruses. In 1980, for example, the WHO declared that the smallpox virus had been eradicated from the world. Are the advances in molecular genetics likely to bring more of this kind of success? Presumably the more that we know about how viruses work, the easier it is to develop the means to combat them?

'That's right,' Crawford replies. 'Viruses are so small that they were among the first entities to have their genomes sequenced. We've learnt a huge amount; we know, for instance, that certain genes have specific roles to play in the way in which viruses transform the cells which they attack. The big hope in virology is for the development of antivirals which work like antibiotics. There has been tremendous success with HIV, for example. We now have something like fifteen drugs which we can use to keep the illness under control; and from the time of the first-generation drugs, these have been developed on the basis of knowing which genes are required for particular functions of the virus, and then blocking them in various ways. This has been very effective.

Almost certainly there will come a time when we can treat viruses with antivirals just as we treat bacterial infections with antibiotics.'

The success which combination treatments have had in keeping HIV under control raises an interesting side issue to do with the nature of the AIDS illness. In 1996, the retrovirologist Peter Duesberg wrote a book called *Inventing the AIDS Virus*, in which he argued that the HIV virus was not causally implicated in the development of AIDS. Given what we now know about the HIV virus, and given the success of the new drug treatments, is this still a sustainable position?

'No, definitely not,' insists Crawford. 'In fact, there was evidence all along that the HIV virus was the cause of AIDS; and even if there was a modicum of doubt at the beginning, the hype which accompanied Duesberg's claims was very unhelpful. What I found most distressing about the whole affair was that people with the disease wanted to believe that it wasn't caused by a virus, because they didn't want to die. But it just wasn't fair to give them false hope in this way. We also felt that the media played a malign role in that they were happy to publish the "AIDS isn't a virus" propaganda, without listening properly to what other virologists were saying. But this has died down now. Really, the success of the combination therapies has made the issue go away.'

Whilst it is well known that viruses are the cause of illnesses such as influenza, AIDS and smallpox, it is maybe less well known that they are also implicated in the development of certain cancers. Perhaps the best example of this is the role which the human papilloma virus (HPV) plays in the aetiology of cervical cancer. By what processes do viruses result in cancer?

'Well, the danger normally comes from those viruses which cause persistent, rather than acute, infections,' Crawford replies. 'The virus gets into the body, and whilst it won't necessarily cause a noticeable problem, the immune system, for one reason or another, is not able to eradicate it. What one often finds then, as I mentioned before, is that these viruses can only replicate themselves if the cell which they occupy is itself dividing; to reproduce itself the virus requires enzymes which are only available during the process of cell replication. This means that some of the viruses have developed mechanisms to push the host cell into a cell-cycle process; and also mechanisms to keep the cell cycling.

'We now know that there are a lot of viruses like this, and we're discovering more all the time. For example, the HPV virus will cause anything from a benign wart to cervical cancer. It normally enters the body via a microscopic crack in the skin. It then infects cells in the basal area of the skin, and it causes them to proliferate; eventually, because the cells proliferate too quickly, you end up with a little cauliflower like wart, or if the growth gets completely out of control, it will develop into a cancer.'

HPV is the most prevalent sexually transmitted disease in the world; some studies estimate that it infects up to 75 per cent of all women. So why does it only cause cancer in a very small minority of these women?

'Primarily because there are different types of HPV, and not all of them cause cancer,' Crawford answers. 'Obviously, it is vitally important for us to find out what the difference is between the different kinds of HPV. We know that they carry genes called E6 and E7, which are very important in combining with cellular proteins to induce the proliferation of cells which may result in cancer. But there is always a host element to the story; some people, presumably for genetic reasons, are more likely to develop cancer than others.

'There are also non-genetic factors which put some people at more risk from these kinds of viruses. For example, a compromised immune system is often implicated in tumour development. There is a lot of interesting data about people with HIV and HPV, which suggests that they are more likely to develop abnormalities of the cervical mucosa than their immunocompetent counterparts.'

The virus which perhaps most characteristically produces tumours in the presence of immunodeficiency is the Epstein–Barr virus (EBV). Particularly, it is associated with a high incidence of lymphomas in people with HIV and people who have taken drugs to suppress their immune system after undergoing organ transplants (immune system suppression after transplant surgery is necessary in order to minimise the chances of organ rejection).

'The first thing to say is that the virus infects about 95 per cent of the population, and normally it doesn't cause too many problems,' says Crawford, when I ask her about EBV. 'Most people get infected in childhood, and they don't experience any symptoms. If a person is infected during adolescence or early adulthood, often they'll get gland-ular fever, but then usually nothing else (although they do carry a slightly higher risk of developing Hodgkin's disease). However, if a person is immunosuppressed, then they might well develop an EBV-associated tumour. This tells us that in healthy people it is the immune response which keeps the virus in check.'

Crawford explains how her work at Edinburgh University is shed-ding light on the role of the immune system in combating EBV. 'A lot of work has been done to determine which part of the immune response keeps the virus in check. We now know that the cytotoxic T cells, or killer T cells, are essential. These get knocked out by immuno-suppression drugs after organ transplantation, for example, which explains why some transplant patients get EBV-associated tumours. What we've been trying to do is to replace the T cells in these patients. Unfortunately, it isn't possible to develop just one line of T cells which can be infused into all individuals. It is necessary, just as it is with an organ transplant, to match the tissue type of the T cell with the tissue type of the patient. So we've developed a huge bank of about one hundred different kinds of T cells, all tissue typed, and all specific for EBV proteins.

'We've just completed a pilot study with eight patients and we've got a very good result. After we infused the T cells, five of the eight got complete tumour remission, despite the fact that they had failed on all other treatments. This is very encouraging. The next step is more trials. We're currently in the middle of a much bigger randomised control trial. It's very early days, but so far things are looking good.'

How long will this trial process likely take?

'If we were developing just a normal drug, then the treatment would become available as soon as the trial was finished (assuming that we got good results). But what we're doing is a bit more complicated. It is probably fair to say that we're at the beginning of the process of developing an easy-to-use treatment. Given a bit of time, we should be able to develop quite a sophisticated approach. Now that it is possible to clone the receptors from T cells, and to put them into other cells, we hope to bypass the requirement for tissue type matching and have one T cell for all. But first we need the basic knowledge that we're doing some good, and that's the stage that we're at.'

When they read about this kind of research, do people suffering from EBV-related tumours ask whether they can be included in the trial?

'Yes, we get enquiries from all around the world,' Crawford replies. 'It's quite distressing. People are often desperate. But I have to point out that we're very much at an early stage with the research; and often it turns out that people don't have the right kind of tumour anyway.'

If it becomes clear during a trial that a team is getting very good results, is it possible to move more quickly through the trialling process, so that people who are desperate can get the treatment?

'Yes, it is,' Crawford answers. 'Every trial has a data monitoring committee; they look at the data as it comes in, and if they determine that there are statistically significant data to show that a treatment is better than the control, then it's only ethical to stop the trial, and give everybody the treatment.'

There is an interesting general issue here about the merits of medical trials. The accepted practice is not to make a treatment available until it has been properly trialled and found to be safe and effective. However, this was challenged late in 2002 by the father of a victim of variant CJD, a fatal neurological condition. He successfully argued in the UK courts that his son should be given an experimental treatment, even though both its efficacy and safety had not been determined. What is Crawford's view about the requirement for rigorous, double-blind testing in medical research?

'In general terms, I think it is needed,' she replies. 'But I also think that the process could be speeded up a lot. The number of hurdles which have to be jumped in order to run one of these trials is quite extraordinary. It could certainly be made easier for the research teams. I may be an expert on Epstein–Barr, but I don't have extensive experience of committees and form filling; it's an unfamiliar world, and it would be nice if they made it more user-friendly. In fact, things are likely to get worse, because the EU is launching in with their recommendations. The whole process is over-complicated, and it wastes a lot of time.'

Although it is undoubtedly the case, as we have seen, that increased understanding of the workings of viruses will bring benefits in terms of our ability to treat a variety of illnesses, it also brings with it a number of new problems. Particularly, there are issues to do with the possibility that genetically modified viruses might be employed as terrorist weapons.

'This is a very real threat,' Crawford says. 'Just think about the flu virus for a minute. If somebody produced and released a new strain of flu, then there is a good chance that we'd get a pandemic as a result. It is possible that it wouldn't kill huge numbers of people, but it would certainly cause massive disruption and lay a lot of people low for a while. But what about if somebody produced something just a little bit more lethal than flu, but which spreads in the same way? That could be devastating.

'If there are people with skills like mine, who are also bioterrorists, then they can do this kind of thing. They could manipulate a virus and produce, for example, some kind of hybrid between Ebola and small-pox, and that would be horrific. It is a real possibility. I've never really been able to understand why people talk so much about anthrax as being the great threat. Although I'm not a bacteriologist, it does seem that it is rather hard to infect large numbers of people with it. But this just wouldn't be the case with a virus; smallpox on its own, for example, could be devastating.'

Perhaps the other major concern facing microbiology at the present time has to do with bacteria, rather than viruses, and the fact that the shelf-life of the current crop of antibiotics might be coming to an end. The problem has been caused by the fact that bacteria have evolved resist-ance to the antibiotics currently in use (in part as a result of the overuse of antibiotics). Is this something that we should be worried about?

'Definitely,' replies Crawford. 'In a way, this is more worrying than something like SARS. Although SARS hit the headlines, it wasn't a big epidemic, and it was very efficiently dealt with. But MRSA, for example, has been in our hospitals for quite some time, and it isn't something we've been able to deal with properly.

'There are various things we should be looking at here. First, we need to control the free sale of antibiotics across the counter, and to regulate which antibiotics are used with which particular diseases. There are good data which show that where you get the free sale of antibiotics, you get more resistance developing in bacteria in the community. Second, we need to look at the issue of animal feed. This is a particular problem in the USA, where the gene pool of bacteria is being affected by the fact that antibiotics are given to animals for no good reason. And, of course, the other thing we have to do is to develop new antibiotics.'

One moral of the story of increasing bacterial resistance is that it is the nature of evolution that we're always likely to be involved in a battle – an arms race, if you like – with infectious agents like bacteria and viruses. Does Crawford see a day when we might win this battle once and for all?

'The first thing to say is that we're always going to be in a catch-up situation with respect to microbes,' she replies. 'Their short replication time means that they can mutate very rapidly, whereas it takes us much longer to adapt to them. HIV, for example, changes very quickly, so although there is an immune response at first, in the end the immune system isn't able to cope.

'In terms of the general point about whether we are ever likely to live in a microbe-free environment, the idea might be that because we've managed to eradicate one virus, in the end we'll eradicate all viruses. But I doubt that this will ever happen; not least there will always be new mutations which will mean that viruses which in the past were not able to infect us, will become infectious.

'Also, it is not entirely clear it would be a good idea to live in a microbe-free environment,' Crawford says. 'For example, there is the possibility that by eradicating a virus you will create a space which other entities will come to fill. So whilst it was a good thing that we got rid of smallpox, there is the chance that monkey pox will take over where smallpox left off. It isn't entirely clear how much sense this idea makes, but it is certainly something we have to think about.'

Aren't there also some radical types who argue that because viruses are a life form they should be preserved?

'Yes, but I don't have any sympathy with that view,' Crawford responds. 'I would have pressed for the complete elimination of the smallpox virus. Of course, whether that would have been a good thing is now complicated by the terrorist issue.'

How much threat are human beings under from microbiological agents of illness? There is an argument that the genetic diversity of the human species will afford us a certain amount of protection against even the most devastating of pandemics.

'Yes, this is my view,' says Crawford. 'In any pandemic, it is only a proportion of the people who get infected who get ill. It's what we call the iceberg effect. With any virus, there will likely be a varying proportion of people that you would never know were infected unless you did an epidemiological study. It is certainly the case with flu, and is massively the case with polio. So yes, I think that human beings are genetically diverse enough to be able to deal with the threat from microorganisms.'

8   On Cancer Research

In conversation with Mike Stratton

Cancer has been a recognised disease for the entirety of recorded history. The earliest known references to it occur in the Edwin Smith and George Ebers papyri, which were found in Egypt, and which date back approximately to 1600 BCE. Since this time, there have been many different ideas about the possible causes of cancer. For example, the ancient Greeks believed that cancer was the result of an excess of black bile, thought then to be one of four body *humors*; in the seventeenth century, on the basis of the work of scientists such as Gaspare Aselli, it was held that an abnormality in lymph – a body fluid which contains white blood cells – was responsible for cancer; and in the late nineteenth century, the trauma theory had it that cancer was caused by bodily injury.

However, it was only really in the twentieth century that progress was made towards a correct understanding of the aetiology of cancer. Perhaps the key insight has been that cancer is at base a genetic disorder; or as Professor Mike Stratton, Head of the Cancer Genome Project at the Wellcome Trust Sanger Centre in Cambridgeshire puts it, cancer is the consummate disease of DNA.

'I mean by this that in contrast to most other diseases *every* cancer is due to abnormalities of DNA,' Stratton tells me. 'Over time, the cells in our bodies inevitably acquire changes in their DNA. Partly, this is because they are being continuously bombarded by external factors, such as chemicals, radiation and viruses; and partly, it is because mistakes very occasionally occur during the process of cell division. Most of the changes which accrue have no effect; the cells function perfectly well. Sometimes, there might be a change which causes a cell to die, but so long as this doesn't happen too often, it doesn't matter. Very occasionally, however, a cell in the body will acquire changes in critical genes, which will then make the cell grow perhaps a little more quickly than it should, or to take up a slightly different position, or perhaps to float off around the bloodstream. In other words, the cell begins to behave like a cancer cell, and it does so because it has acquired abnormalities in its DNA sequence in the appropriate set of genes. All cancers have this cause. Whatever factors might affect the incidence of cancers, they all ultimately work through the acquisition of these abnormalities in DNA.'

An interesting point about this process is that cells which are functioning normally have built-in mechanisms which prevent them from replicating in an out-of-control manner. Is it the case then that in those cells which become cancerous there are multiple genetic abnormalities?

'That's right,' Stratton says. 'Part of the way that cells protect themselves is to erect several hurdles which have to be jumped before they will exhibit characteristics of cancer. What we think happens is that an individual cell may acquire one of these cancer-causing mutations, and it may then proliferate a little bit out of its normal proportions. We've probably got hundreds or thousands of cells like this in our bodies at any one time, and they're not causing any problems. But amongst these, there might be one which acquires a second mutation, which will cause it to proliferate further. And this time it may result in a little benign tumour, a clonal expansion of cells which have acquired abnormalities in the cancer-causing genes. The key point is that it is only when the requisite number of abnormalities have been acquired that the cell will manifest the characteristics which we recognise as being a cancer; that is, it becomes invasive and metastatic. At the moment we don't know what this requisite number is; five or six would be a reasonable guess, but it is possibly as many as ten or twenty.'

Does the number of abnormalities required vary between different kinds of cancer?

'Yes,' Stratton replies. 'There seem to be certain cancers – for example, some breast cancers – which have many abnormalities, and others – for example, leukaemia – which are much quieter. Our general view is that the quieter cancers will probably have one or two abnormalities driving them, whereas other cancers – for example, the common epithelial cancers – will be driven by as many as perhaps ten abnormalities.'

There are some cancers which are much more common than others; for example, breast cancer is relatively common, whereas islet cell cancer is relatively rare. Does this mean that cancer-causing mutations are naturally more likely to occur in some cells than they are in others, or is it that external influences are more pernicious in the case of the common cancers?

'This is a very difficult question to answer,' Stratton replies. 'I know this might sound silly, but we don't really understand what makes some cancers more common than others. If we consider, almost from first principles, which organs of the body ought to have a lot of cancer, and which ought to have very little cancer, we immediately run into difficulties. For example, take a three-way comparison between the breast, colon and brain. Breast cancer in women is about three times as common as colorectal cancer, and about twenty times as common as brain cancer.

'The first thing you might think is that cancer will be more likely in an organ with a lot of cells. However, there are something like a hundred times fewer cells in the breast of the kind which give rise to breast cancer than there are cells in the brain of the kind which give rise to brain cancer. So immediately there's a paradox. Breast cancer is relatively common; brain cancer is relatively rare.

'So perhaps we can make this more sophisticated. Maybe it is those cells which normally divide a lot which are most likely to become cancerous. For example, we might expect active cells to acquire a greater number of abnormalities than less active cells. So how does this go with our breast, colon and brain comparison? It is true that the cells in the brain don't divide very much after birth; they're pretty quiet. But the colon is a different matter. The colon lining – the epithelium – replaces itself every twenty-four to forty-eight hours; these cells are continuously dividing, much more so than the cells in the breast. But breast cancer is more common than colorectal cancer, so again we're left with a paradox. Clearly then there is something deeper going on, either in terms of the exposures to which these cells are subject, or in the biology of these cells, which accounts for the differences in the incidence of cancer.

'But the key piece of data, which we find almost impossible to explain, has to do with where the ileum, the last part of the small intestine, joins the secum, the first part of the large intestine. The ileum turns over very rapidly, if anything more rapidly than the colon. The contents of the ileum are no different from the contents of the first part of the colon. The colon has got a high rate of colorectal cancer; the ileum has an incredibly low rate. So there are two kinds of tissue which are adjacent to one another: they're both turning over quickly, they're both big organs, they both contain the same combination of fluids and mutagens, and yet a distance of 5 centimetres makes the difference between a region where you hardly ever see cancer and a region where it is very common.

'It just doesn't seem possible to explain this in terms of cell numbers, cell turnover, the number of mutagens, and so on; it has to be something very deep in the biology of these cells which explains the different rates of cancer.'

Cancer then is fundamentally to do with abnormalities of DNA. But this is not to say that any particular cancer is inherited. For example, it is thought that only 5 to 10 per cent of breast cancers are inherited.

'That's right,' Stratton confirms. 'There is, of course, a risk that *certain* cancers will be inherited; and there are some specific abnormalities in genes, which, when transmitted from parent to offspring in the normal way, almost inevitably mean that the person carrying them will get cancer. But the important point is that even where a cancer is linked to no inherited predisposition, where it is just a matter of bad luck, it is still caused by DNA abnormalities; it is just that they have been acquired since conception.'

Presumably then it is this fact that all cancers are driven by DNA abnormalities which motivated the setting up of the Cancer Genome Project in 1999?

'Well, actually, we've known for nearly a hundred years, in general terms at least, that cancer is linked to abnormalities in genes,' Stratton says. 'But with the advent of recombinant DNA technology, it became possible to identify some of the key genes, which, when abnormal, drive cells to become cancer cells. However, to find these genes, we always required a prior clue, so that we could determine whereabouts in the genome we should look. However, the problem is that there is no reason why every cancer gene should provide us with a positional clue. It could be buried somewhere deep within the genome, where it might never be found.

'It was this which motivated the Cancer Genome Project. When we set it up, we envisaged that rather than waiting for positional clues, we could go through the genome, gene by gene, and ask the question: Is this gene abnormal in any of the variety of human cancers? If the answer is no, then it isn't a cancer gene. In this general way, it should be possible to interrogate every gene for every kind of cancer. Although we can't do it straightforwardly, this is the conceptual underpinning of the Cancer Genome Project.'

There has already been some success in the treatment of cancer which has resulted from increased understanding of the functioning of genes. Perhaps the most startling example is the case of the drug Glivec.

'Glivec is a perfect example of the way that many researchers thought that recombinant DNA technology could be employed to help find treatments for cancer,' Stratton explains. 'It was felt that if it was possible to find the genes which drive cancer, then there was the potential for developing drugs to switch them off, thereby switching the cancer off. This is a very simple, but powerful, notion. However, it was greeted with some scepticism by the clinicians who treat cancer; they probably thought that cancers were a lot more complicated than these rather brash, young molecular biologists appreciated. I suppose to some extent, prior to Glivec, their scepticism would have seemed to have been justified. A reasonable number of cancer genes had been identified, yet this had not led to clear therapies to combat cancer. Indeed, there might even have been some loss of faith amongst the molecular biologists themselves. But Glivec is a story of how this strategy worked extremely well.

'Chronic myeloid leukaemia (CML) is a disease in which in almost every case there is a rearrangement between chromosomes 9 and 22. Roughly speaking, genes on both chromosomes break away and are brought together to form a new gene. It is this rearranged gene which drives the leukaemia; a protein encoded by it is constitutively switched on, with the result that it sends signals into the cell telling it to grow, which is what it does. So there was an attempt to develop inhibitors of the protein which is encoded by this rearranged gene.

'I remember quite clearly the first time that I was told about what had been achieved. I was at a conference in Australia, quietly snoozing my way through it. The presentation was quite cautious, but when it was said that some 33 out of 34 patients, who had all failed all other kinds of treatments, had gone into remission with this inhibitor, I was wide awake; you just had to sit up and take notice.

'And indeed, the success of Glivec in treating chronic phase CML has been quite extraordinary – much better than any of the brash, young molecular biologists from twenty years earlier would have dared to fantasise about. It has transformed the world of cancer research. It means that people now recognise that by switching off a gene, it might be possible to switch off a cancer.'

Stratton has noted in the past that cancer cells are 'very clever guerrilla fighters', and that it would not be unexpected if they found ways to combat drugs like Glivec. What's the thought here? Is it simply that there might be a few cells within a tumour which will have mutated in such a way that they will have the characteristics which enable them to resist a particular drug; and that this will mean that as soon as the drug is administered they will enjoy a selective advantage over the other cells in the tumour, and therefore come to dominate it?

'Yes, I think there is evidence for something like this,' Stratton confirms. 'As a cancer grows and divides, it inevitably generates mutations; the precise number will depend on how long the cancer has been growing and the rate at which mutations occur. If something like Glivec is then applied to it, it is conceivable that the vast majority of cells will be killed, but that there will be a very small minority that have abnormalities, which previously didn't provide them with any advantage, but which allow the cancer to survive in the context of Glivec. This is the most plausible model for how resistance to Glivec, and probably other treatments, arises.'

Does this mean then that we need to look for cancer treatments which work in the same way as the combination therapies for AIDS; that we need to catch cancer early, and then to change the kind of drug which is being administered once any cells with inbuilt resistance begin to make their presence felt?

'Well, this kind of model actually originates with cancer treatment,' Stratton replies. 'Many old-style cancer treatments are combination therapies; the patient is given two or three different drugs. It isn't always clear why two or three drugs should work better than one. But, in the context of the model which Glivec suggests, it is possible to see how this might happen. If Glivec kills off all the cells in a cancer except for say ten, and another drug also kills off all the cells in a cancer except for ten, the probability that there is a cell amongst either group of ten which has both resistances is very small. So, in this way, there is an increased chance of killing the whole cancer. Indeed, people are trying to develop multimodal, combination therapies, even using these new agents like Glivec, in an attempt to fight off the ultimate threat posed by cancer, which lies in the fact that, as a clever guerrilla fighter, it generates resistance.'

One of the issues which a drug like Glivec brings to the fore has to do with the economics of developing treatments which target specific cancers. Particularly, it isn't clear how pharmaceutical companies can make money by producing drugs which might only be useful to a small number of people. Is this something we should be worried about?

'The general problem is as you describe,' Stratton says. 'Putting it simply, from a commercial point of view, pharmaceutical companies will be most attracted to drugs which target illnesses which are common, chronic, and preferably not fatal. The trouble is that cancer doesn't really fit in these categories. Cancer is a common disease, if we're talking about everything which comes under the general heading of cancer. But it is possible to divide up cancer in various ways. For example, if we're talking about the organs within which a cancer originates, then cancer of the breast behaves differently from cancer of the colon or brain cancer. But we can also subdivide the cancers which occur *within* particular organs; the more we learn about breast cancer, for example, the more subdivisions we find. If we then get to a situation where we're identifying abnormal genes, we may divide up cancer into a very large number of categories; for example, there could be twenty or thirty different kinds of breast cancer alone.

'It is fairly clear that we're talking here about a situation which doesn't provide the pharmaceutical companies with targets which are obviously commercially viable. So there is this central uncertainty about drug development for cancer treatment: just how fragmented will the disease become?

'There are a number of points to be made about this. The first is that it is likely that non-commercial organisations will become more involved in the early stages of the process of finding drugs which target particular molecules. Indeed, this is already happening; as the research process has become cheaper and better understood, cancer charities and government-funded research teams have begun to take on the initial screening for molecules.

'However, things might become more problematic after this stage. It is just possible that pharmaceutical companies will be willing to become involved in the later stages of molecular tinkering, if they think that what they're working on is likely to become a good treatment. However, this would still involve them in large costs, so it isn't by any means certain. In short, it still isn't clear whether the development of novel cancer drugs, in a world where the disease is fragmenting into different subcategories, will be commercially viable; and if it isn't commercially viable, it isn't clear who will take on the development.'

The other debate which occurs at the intersection of commercial and research interests has to do with the patenting of genes. Roughly speaking, the issue splits between those people who argue that gene patenting is necessary to provide an incentive for pharmaceutical companies to engage in research which can be very expensive, and those who argue that knowledge about genes is not the kind of thing which should be patentable, and that gene patenting is likely to hinder the flow of information and to make the potential benefits of gene technology subject to a profit motive. What is Stratton's take on this debate?

'Well, in order to think about this issue properly, one has to stand back a little bit from the patenting of genes, to think about the benefits of patenting generally,' he answers. 'Patents are designed to encourage innovation and investment. So, for example, if one looks at the kind of pathway from gene to drug we've described for Glivec, whilst there wasn't, and isn't, any need to motivate people to look for the genes which these sorts of drugs target – we are all very keen to do this – incentives probably are needed to persuade drug companies that it is worth spending the huge sums of money which are required – to satisfy regulatory requirements, and so on – to bring a drug like Glivec to market. In this sense, then, drug patents are generally, though not always, socially beneficial.

'It is true that once a company has patented a drug, then they can charge anything they like for it. Market forces mean that companies tend to be more or less realistic about pricing in the Western world. However, there are difficulties, of course, when one is thinking about other parts of the world, where often there isn't the money to pay for particular drugs. This has required some rethinking about how patents are implemented, and we now have something like a two-tiered approach, which sees the Western world paying a certain amount for drugs, and other parts of the world paying much less, and sometimes nothing at all.

'Despite all the furore about antiretroviral drugs and the Third World, there is certainly an argument to be made that if it weren't for patents, then there might not be antiretroviral drugs. So it is necessary to look at this issue with considerable care and intelligence. It isn't something about which one should take knee-jerk political stances.

'However, in the case of genes, the situation isn't as clear as it is for patents on drugs,' Stratton says. 'It isn't obvious that patenting genes for their utility has benefits for humankind. For a start, people will always be willing to look for the uses of genes, regardless of whether there are patents involved. The proponents of patents would say that patents bring in more investment, and that therefore more gene usages will be discovered. But even if this were true – and, in fact, the benefits are probably fairly limited – it would be necessary to balance it against the problems which arise from patent monopolies. We can see the kinds of problems which arise if we look at something like the breast cancer genes. Health services cannot use these gene sequences in a completely free way; they have to pay licence fees which are sometimes expensive, and often they cannot do diagnostic tests themselves at all – they have to be sent out to be done elsewhere.'

It is interesting that Stratton mentions here the case of breast cancer genes, because he headed up the team which patented BRCA2, one of two breast cancer genes, both discovered in 1994, which account for a large proportion of inherited breast cancer.

'My motive for allowing that patent to be taken out was largely that it was a defensive position we could adopt,' he explains. 'To some extent, this has been justified; there have been attempts to monopolise the gene, and although it isn't the case that our patents are an incredibly strong defence against this, they are probably the only weapons in the toolbox. In a world where patents exist, you have to make a decision whether you're willing to fight for control of your gene or not.

'But we don't ask for any money from people who are using the gene,' he emphasises. 'My view is that since BRCA2 was discovered in the context of publicly funded research, being conducted in order to improve human health, we should continue to facilitate improvements in human health by making this sequence available, as far as is possible, for people to use.

'However, if this sequence had proved to be so difficult or expensive to use that nobody was using it, and, therefore, no diagnostic or pre-symptomatic testing for breast cancer was taking place, then it would have been very important to have taken out a patent, and to have licensed it to a commercial enterprise. This would have enabled the enterprise to have brought in investment capital, and to have offered a diagnostic service. Obviously, it would have been very expensive, but it would have been better than no service at all. So again, it is important not to jump very quickly to politically led conclusions. Nevertheless, in terms of the balance of the issue of patenting, I think that most of the gene patents probably do not benefit humankind, even though it is important to make this judgement on a case-by-case basis.'

When talking about the significance of the Human Genome Project most scientists now strike a fairly cautious note about when we can expect to see major breakthroughs in terms of human health and life expectancy. What's Stratton's view about how quickly we're likely to see progress in cancer therapies?

'Well, there are enough uncertainties in the field of cancer research that bold predictions about the future are probably not very sensible,' he replies. 'But having said that, there are a number of different ways we can think about what might be likely to happen in the future. For example, it seems probable that as we know more about the machinery of cells, we'll be in a much better position to find out what goes wrong inside them and how to put this right. Also, since cancer is a disease of DNA, it is likely that any increases in knowledge in this area will lead to new avenues of research and possibly new treatments. In this regard, the advent of Glivec, and all that is implicit in this discovery, is a cause for considerable optimism.

'However, there are many things we can't predict, so cautious optimism is probably the best outlook. We don't know, as we have mentioned, how resistance to drugs will play out. It is possible, for example, that Glivec may be very successful for four or five years, and then simply cease to work. We don't know yet, and we won't until the studies have been done. We also don't know how the paradigm of Glivec will extend to other kinds of cancers, especially the common epithelial cancers. We can't think of an obvious reason why it wouldn't extend, but cancers are capable of springing surprises, so there are no certainties. Undoubtedly, the world of cancer research has been transformed by the success of Glivec, but the very fact that it has been so transformed gives one the sense that it might always be transformed back again.

'The other point, of course, is that we're not going to get proper outputs from the human genome for something like another ten to twenty years. I have frequently said that I would be surprised if cancer treatments weren't very different in twenty years from now. But I always add the rider that it will be interesting to examine this statement in twenty years' time to see how things have played out, how much hasn't been fulfilled, and to see the reasons for this.'

Presumably the chances are that there won't ever be a single magic-pill cure for all cancers, but rather cancer is likely to become increasingly a chronic manageable disease, perhaps similar to something like diabetes?

'That is certainly one vision of the future,' Stratton says. 'One single magic-pill is unlikely. We have already mentioned the fragmentation of cancer into many different diseases, such that each disease will require its own pill or set of pills. Will it be management, keeping it in a chronic phase, rather than cure? Well, this has been said, and it is being said more and more frequently. Obviously, it would be a good outcome, and it certainly provides us with a different strategy for dealing with cancer. However, I'm really not sure that this, rather than treatments which completely eradicate cancer, is the most likely outcome. My suspicion is that in fifty years' time, there will be a mixture of both these things going on.

'Implicit in making this kind of judgement about likely outcomes is knowledge of how cancers behave. A lot of cancers do simply explode when you give them drug treatments. For example, even old-style drug treatments result in a 95 per cent cure rate for testicular cancer; the cancer doesn't come back, it's not being controlled, it's being cured. We tend to forget this. Is that what we're going to see with the common epithelial cancers, or are we going to see a set of strategies aimed at holding them in check? I actually wouldn't be bold enough to answer this question. At the moment, I really don't know.'

9   Animal Experimentation, Ethics and Medical Research

In conversation with Colin Blakemore

In May 2004, the British government announced that it was establishing a new national centre which would fund work directed towards the aim of replacing, refining and reducing the use of animals in scientific research. Lord Sainsbury, the Parliamentary Under-Secretary of State for Science and Innovation, marked the announcement by insisting that whatever the ultimate aim of the centre, it was clear that animal testing was currently necessary for proper scientific research. In this, he was echoing a statement which he had made some months earlier, reaffirming the commitment of the British government to animal research undertaken within existing regulatory frameworks.

However, this claim that animal experimentation is necessary for scientific research is not universally accepted. Perhaps not surprisingly, it is most keenly contested by the animal rights lobby. The National Anti-Vivisection Society, for instance, argues that 'the fundamental flaw of animal-based research is that each species responds differently to drugs and chemicals, therefore results from animal tests are unreliable as a means of predicting likely effects in humans. Thus, animal experiments are unreliable, unethical, and unnecessary.'

Whilst this view might justifiably be regarded as being rooted in the hyperbole of a pressure group, the animal rights lobby is not alone in suggesting that the issues surrounding animal experimentation are difficult. For example, in a *BMJ* article titled 'Where is the evidence that animal research benefits humans?' (28 February 2004), Pandora Pound *et al.* noted in their first sentence that: 'Clinicians and the public often consider it axiomatic that animal research has contributed to the treatment of human disease, yet little evidence is available to support this view.' As a result of this claim, the British media reported that scientists were beginning to doubt whether animal research was useful at all.

'It was completely predictable that that phrase would become a kind of rallying call for the animal rights movement, and it is very unfortunate,' says Professor Colin Blakemore, Chief Executive of the Medical Research Council, when I mention the article to him. 'But it is possible to counter the claim with the fact that the Royal Society has recently published a pretty comprehensive study which says that virtually every medical advance in the last century has depended on the use of animals in research at some point; or with the statement from the Department of Health, in its evidence to the House of Lords Committee on Animal Experimentation, that the National Health Service could not operate without the foundation of the knowledge which animal research has built. The overwhelming view of the scientific establishment, and I'm not using that expression in a pejorative way, is that animal research is necessary for progress in medical science.

'Also, it is important to point out that there has been some misinterpretation of the data which Pound *et al.* analyse in their paper. They found, and I think they're right about this, that the transition between animal research and clinical application often lacks rigour in terms of a proper review of evidence. The eagerness of some scientists or drug companies to try out new treatments can lead to the misinterpretation or over-interpretation of data from animal research. But they only examined six areas of research, out of the many hundreds where it is well documented that animal experimentation led to clinical applications. Specifically, they looked at six comprehensive reviews of animal evidence and the associated attempts at clinical application.

'In one of these cases, they found that despite clear evidence from the work of people like Michael Marmot that human health is linked to social status and responsibility, there is no analogue of this phenomenon amongst primates. But this isn't surprising; it doesn't work this way in primates. Marmot was looking at the Civil Service, which is not exactly the kind of environment you can easily imagine being mimicked in primate groups.

'The other five cases were situations where clinical trials, or the preliminary treatment of human beings, had started on the basis of a cursory examination and often very optimistic interpretation of previous animal research. In every case, it turned out that the clinical technique did not work. In retrospect, when the animal research was thoroughly and systematically reviewed, it turned out that the results from the animal experiments didn't justify the clinical application. So far from showing a mismatch between animal and human results, it showed a perfect correlation. The failure was in the transition between the animal results and the clinical studies.'

What then of the criticism that the small differences in the makeup of complex systems such as animals and humans can make a lot of difference in terms of how they respond to chemicals and drugs, and that, as a result, extrapolation from animals to humans is necessarily unreliable?

'There are probably only two or three properly qualified people in the world who hold this position,' Blakemore replies. 'Ninety-nine per cent of physicians in the United States say that it is essential to use animals in medical research, and more than 95 per cent of British physicians say the same thing. So whilst it is important to listen to maverick opinion, it is clear we shouldn't put too much weight on it when one considers that the American Medical Association, the Royal Society, the British Medical Association and the General Medical Council all state that animal experimentation is necessary.'

Presumably then it must be possible to give specific examples of the importance of animal research which will settle the issue?

'Yes, almost everything is an example,' says Blakemore. 'Every drug, every form of advanced surgery, nearly every antibiotic, nearly every vaccine – the development of all of them has at some stage involved animal testing. What more evidence is there? By law, every drug has to be tested on animals before it can be used on humans; you cannot get a prescription for a drug which has been developed in the last 100 years which hasn't been tested on animals. So when you ask for examples, just about everything is an example: every pain killer; every treatment for heart disease, kidney disease and cancer; chemotherapy; radiotherapy; surgical techniques; bypass surgery; open heart surgery – you name it, animals were involved in the research.'

So what are the motivations of the people who argue that animal research is unreliable and unnecessary? Are they just pursuing a particular moral agenda?

'I'm not sure,' Blakemore answers. 'I know that Ray Greek, for example, has argued that his is not a moral position; he doesn't object to the use of animals in medical research on moral grounds. He has said, much to the consternation of his own supporters, that if it took the death of 10,000 chimpanzees to find a cure for AIDS, then he would be in favour of it. His argument then is a factual one: that animal research doesn't work. But I would be extremely critical of the evidence he cites to support his view. Mainly he makes use of excised quotations, which seem to suggest that eminent researchers doubt the value of animal research. But if you go back to the source of the quotations, you find that they are being used out of context, and that the thrust of the papers they appear in contradicts the argument the quotations are being used to support. I don't call this evidence.'

Perhaps the area of animal research which causes most controversy is that which involves primates. In the UK, this issue became headline news at the beginning of 2004, with the announcement by Cambridge University that it had axed plans to build a new multi-million-pound centre for primate research. The BBC reported that the decision to abandon the project was based in part on the spiralling cost of satisfying the requirements of animal welfare legislation and the need for security to protect against the threat of attack from animal rights activists. The decision was greeted with regret by many people in the scientific community. For example, Mark Matfield, Director of the Research Defence Society, called it a 'serious blow for British medical research' and argued that 'the government needs to bring in tougher legislation to tackle extremist campaigns, otherwise they will remain a threat to all medical science that depends on animal research.'

What is the importance of primate research? It accounts for only a tiny percentage of the total of animal research, so presumably there are some quite specific reasons for doing it?

'Yes, that's right,' Blakemore confirms. 'To get a licence for primate research requires that a very special case be made: not only that animals are necessary for the research, but also that it must be primates, that no other animal will do. Moreover, particular attention is paid to the question of suffering; it is much harder to get a licence where primates might suffer than it is in the case of say mice. Also, the number of animals to be used and the objectives of the research will be taken into account. What this means is that research with primates will be of great strategic importance in terms of its potential to deliver results, and it will be work which simply cannot be done with other species. We're talking here primarily about three main areas of research: endocrinology, where for many of the hormone systems, the monkey is the only model which has the required likeness to human beings; neuroscience, because of the similarity between the organisation of the primate and human brain; and some areas of vaccinology, particularly, for example, in attempts to develop AIDS vaccines.'

The specific worry about primates has largely to do with their cognitive abilities; particularly, there is the possibility that they are self-aware; self-conscious, rather than simply conscious. The philosopher Peter Singer, for example, in his book *In Defence of Animals*, asks why we are willing to lock up chimpanzees in primate research centres and subject them to painful experiments when we would never think of doing the same thing to a retarded human being who had lesser mental abilities. Singer concludes that it is only speciesism which explains this difference. What does Blakemore make of this kind of argument?

'I know the arguments, and would keep an open mind on the nature of the evidence about something like primate self-awareness,' he replies. 'But I think it is very important to hang on to one strong moral principle, which is that there is a clear distinction between our responsibilities to our own species and our responsibilities to other species. Most people, given the choice between saving the life of a human or an animal, will think that their primary obligation is to the human being. I'm very fond of animals, I have kept pets all my life, but if it came to a choice between my cat, which has lived with us for some seven years and is very much a part of the family, and the life of one of my daughters, I would not have the slightest hesitation in saying that the life of my daughter should have priority; and I think that most people would feel the same way. They might love animals, but they see that human beings are just different.'

The response which somebody sympathetic to the Singer view would likely offer to this argument is that at least part of what leads us to the thought that humans are different from animals might be present in a primate and absent in a severely disabled human being.

'Yes, the idea that all that matters is sentience,' says Blakemore. 'But we need a firmer foundation than this to base our judgements upon. In the end, all it amounts to is an anthropomorphic claim about what it must be like to be a monkey. However, this can lead to serious errors of judgement, such as that of Peter Singer himself, when he argues that there is a line to be drawn in terms of sentience somewhere between rats and fish. But why? Just because we can't get ourselves into the mental life of a fish doesn't mean that a fish is not sentient. The correct starting position is that it is *possible* that all living animals with a nervous system have some kind of experience. Therefore, we have a responsibility to all species to minimise suffering, but on top of this we have a *primary* obligation to our own species. This is a very normal, biological principle. You see it in virtually every species; they treat their own species differently from other species.'

So when Singer talks about speciesism, would Blakemore be happy to accept that as his position?

'Yes, that is exactly my position,' he confirms. 'But I don't accept that "speciesism" is a pejorative term; I'm quite happy to defend the position. In the end, you have to draw a line; you couldn't walk down the street if you really believed that to kill any living thing was a sin or immoral because you'd be worried about the small insects under your feet. There is an extreme version of Buddhism which holds to that position, but taken to its logical conclusion it is ridiculous; you'd end up never moving because you'd be worried about hurting microbes. So it is necessary to draw the line somewhere. The only firm line on genetic and morphological grounds is between our own species and other species. But the fact that we're justified in treating our own species differently *doesn't* mean that we have a kind of Cartesian licence to treat other species in any way that we want; because of our own moral status, we have an extended moral obligation towards the rest of the world. It is just that this moral obligation doesn't entail that we have the *same* responsibilities to other species as we do to each other. So I'm a speciesist, and I would defend that position.'

At the end of 2003, it was reported that Blakemore had been turned down for a knighthood because of his support for vivisection, and that, as a result, he was considering his position as the head of the Medical Research Council. 'It has nothing to do with whether I particularly deserve an honour, that is neither here nor there,' he told BBC Radio 4's *Today* programme. 'The mission statement of the medical research organisation which I now run includes a specific commitment to engaging with the public on issues in medical research. How can I now, in the present circumstances, go to MRC scientists and ask them to take the risk of being willing to talk about animal experimentation with this indication that doing so will reduce their standing and their reputation in the eyes of the government?'

Was he reassured by the government's response to this situation that it fully accepted the need for animal research, and that it admired and supported those scientists who had been on the front line in the struggle against animal rights extremists?

'Yes, I was,' he replies. 'But actually, what was almost more important was the widespread support that came from the scientific community and the media. The support from the media, in particular, was quite extraordinary and a big surprise; virtually the entire spectrum made strong statements about the importance of animal experimentation. So the debate served a useful purpose; it produced a kind of national solidarity, which was much needed. This is also reflected in public opinion. The latest opinion poll shows 90 per cent of the population in support of animal research. It is significant that there is no other major issue where you get this kind of consensus; we still treat the issue of animal research as if it is highly controversial, as if the public haven't made up their mind, but they have made up their mind.'

The opinion poll which Blakemore refers to here was couched in a particular way: specifically, people were asked for their opinion about animal research on the assumption that certain criteria had been met. For example, one question asked whether people could accept animal research for medical purposes, where there was no other alternative. But, of course, it is precisely the claim of the animal rights lobby that there *are* alternatives to animal research.

'Well, if there are, let's see them delivered by those people who claim that there are,' Blakemore responds, when I put this to him. 'I have faced the whole range of arguments from those who are opposed to animal research. I have enormous respect for people who simply say that they don't care about the range of benefits which are the result of animal experimentation; they don't deny that there have been these benefits, but they don't want any part of them because they think that animal experimentation is wrong. It is very difficult to maintain this position, because we all do well as a result of advances in medical techniques; for example, we all benefit from the fact that people are vaccinated, and from our knowledge of the importance of public health. But I understand and respect this position.

'However, I have very little respect for people who say that animals are so very different from human beings that animal research has no relevance for understanding humans; or that all treatments which have been developed on animals are dangerous to humans; or that, animal researchers enjoy what they're doing, that they're basically sadists, and that, anyway, it is only really about filling the pockets of the drug companies. This is not a parody of the kinds of arguments which are made, and I have no time for them. They are rationally indefensible. If there are alternatives, let's see them. We want them. I don't know of a single person who uses animals in their research who wouldn't rather use an alternative. Moreover, there is a great paradox here: the alternatives to animal research which do exist have been developed by researchers who have previously experimented on animals. I've had grants to develop alternatives, I've done a lot of work on tissue cultures, and I use computer simulations for a lot of my work, yet I'm accused of being a villain because I've also experimented on animals.'

The issue of animal experimentation is not the only area of dispute in the field of medical science. There is also, for example, considerable public debate about the use of stem cells – roughly speaking, cells which have the capacity to differentiate into cells of any type – in medical research. What, I ask Blakemore, is the importance of stem cell research?

'It's very easy to slip into hype when answering this question, but even so, I would say that the discovery of stem cells, and of their potential to transform into any tissue type in the body, really offers the possibility of one of the most significant advances in the history of medical treatments,' he replies. 'Most importantly, the kinds of diseases which potentially can be treated with stem cell therapy are a whole range of presently incurable diseases which are growing in rate in the population because of ageing. These are the degenerative diseases. The possibility of replacing tissue which no longer functions properly is extremely exciting. The kinds of conditions which might well be treatable, and for which there are already animal models of treatment, include Parkinson's disease, Alzheimer's disease, diabetes, heart disease, bone disease, stroke and cancer.'

In May 2004, the world's first stem cell bank, partly funded by the Medical Research Council, was opened in the UK. On its opening, it contained two stem cell lines, developed by teams from Newcastle and London. What's the idea behind banking different stem cell lines like this?

'There are actually a number of reasons for banking stem cell lines,' Blakemore answers. 'The main one is to provide uniform standards for typing the cells; for understanding their composition and makeup. It is estimated that perhaps as few as 100 or 200 different stem cell lines, each with different immunological characteristics, will be enough to provide tissue type matching for virtually every person who needs a transplant. So it's a bit like matching blood groups, or indeed tissue typing for organ transplants; it is necessary to get the match right, which is the main reason for collecting more than one line.

'The other reason is that different stem cell lines might have a better capacity to transform into particular kinds of adult tissue. We don't know enough yet about the basic biology of stem cells to know this for certain, but it is possible that although all stem cells from embryos are derived at a similar stage of development, between 0 and 14 days of age, some of the lines will be better at transforming into say heart or brain than others. So this is a second reason for having more than one line.'

The moral objections to stem cell research tend to be voiced primarily by people who object to the use of embryos in the collection of stem cells. For example, Patrick Cusworth, spokesperson for the anti-abortion charity Life, claimed that the use of embryos in stem cell research 'reduces human life to little more than a pharmaceutical product'. What does Blakemore make of this kind of argument?

'Well, I said earlier that we should have a special attitude towards other humans, so crucial to this argument is how we define a person,' he replies. 'A pro-Life group will say that a fertilised egg is a human being. I don't accept this; if any cell which has a full complement of DNA and is capable of transforming into a human being is a person, then every cell in a human body is a person. We now know from Dolly the sheep that if we take the nucleus from an adult somatic cell – a cell from our skin, for example – and put it into a vacant egg, then it can become a human being. Should we therefore be worried about every cell which sloughs off from the surface of our skin? Should we treat each one of those cells as if it were a person? Obviously not. We have to recognise that an embryo, certainly before the nervous system begins to form, is just a bundle of cells.'

People who are opposed to research which uses stem cells taken from embryos also claim that there is no scientific case for the practice; that adult stem cells will do just as well. Is there any truth to this claim?

'This sounds very like the argument which the animal rights lobby makes about the alternatives to animal experimentation,' Blakemore replies. 'My response is to say show me the heart surgery which is going on using adult stem cells. It doesn't exist. It might be that harvesting adult stem cells *will* be the right thing to do at some point in the future. One of the great advantages of adult stem cells is that you can do autotransplantation; you can take cells from a patient, let's say bone marrow stem cells, or neural stem cells from the nose, grow them up so that there are enough of them, and then use them to treat whatever problem the person you took them from is suffering. In this situation, there is no difficulty with tissue typing, it will match perfectly. But we don't have the expertise at the moment to do this; we would need to know a lot more about stem cells, and this knowledge is going to come from the study of embryonic stem cells. But the hope is that one day we will be able to use adult stem cells.'

What about if a person has an illness which is caused by a genetic mutation? Presumably it won't be easy to treat something like Huntington's, for example, with a patient's own stem cells?

'It's true that if a person carries a genetic mutation then it is present in their stem cells; but actually this doesn't necessarily mean that you couldn't use their own stems cells in a treatment,' Blakemore says. 'Huntington's disease is a dominant genetic disorder. The symptoms, though, don't appear until the person is 35 or 40 years old. So although the gene is carried from birth, it doesn't express itself until later; the problems accumulate gradually. It is conceivable, therefore, that if you were to derive neural stem cells from someone with Huntington's disease, and you then transformed them into new nerve cells, they too would last for many years before the condition showed itself. So there is this possibility.

'One of the things which should be emphasised about stem cell research is that it is going to provide much greater insight into the nature of disease. It isn't just a matter of developing new treatments, it is a way to understanding disease. For example, stem cells are helping us to understand the cancer process, because, like cancer cells, stem cells are immortal. Also, our understanding of the normal development of human embryos will be helped by understanding stem cells; we'll learn about the rules which cause uncommitted cells to transform into committed cells. Finally, we can learn about the nature of certain diseases by studying the stem cells of people who carry genetic defects.'

For those people who are suffering from diseases which might be helped by the techniques derived from things like stem cell research, the whole process of medical research can seem frustratingly slow. To a large extent this is inevitable; scientific progress doesn't come easily. However, there is an argument that medical research is being hindered by a growing and intractable bureaucracy. Does Blakemore have any sympathy with this view?

'Yes, I think it is a real problem,' he answers. 'As you suggest, we are in the process of saddling medical research with an unmanageable bureaucracy, each component of which looks very reasonable on its own, but which when collected together becomes a ridiculous burden on medical researchers. For example, many people are forecasting that the new European directive on clinical trials, by introducing greater regulatory controls, will significantly reduce the number of new medicines being developed. It is also possible that the Human Tissue Bill which is going through the UK Parliament at the moment will make it very difficult for scientists to use the information which is gained through routine and anonymous diagnostic tests. The origins of the Bill, of course, are the terrible events which occurred at Bristol Royal Infirmary and Alder Hey. However, in those cases we are talking about the heads of babies being left in fridges for years, which clearly was awful; but it would be ridiculous to think that as a result we should not be able to use urine or saliva which has been collected in anonymous public health screenings. I hope that the government get it right. There is a lot of discussion going on at the moment. The problem is that we tend to go overboard in terms of the extent of the regulation which is introduced. I support the form of the law in this country, but the way it has been implemented has become a huge burden on researchers, and it is impeding medical research.'

Does Blakemore have a view about how the various issues and debates surrounding medical research are likely to play out? Is he optimistic about the future?

'I think it is necessary to learn the lessons of history,' he replies. 'I remember that after the first heart transplant there was intense criticism from the church, civil libertarians, medical ethicists and the animal rights lobby. People complained that the original research had been done on pigs; that the procedure was a dangerous experimental technique; and that it was possible that patients were being coerced into undergoing transplants. All the same kinds of arguments that we hear now were being played out then. But people today think that heart transplants are a medical miracle. The public completely accepts the procedure. Social and moral attitudes then are not absolute and fixed; they are influenced by the evidence of benefit. My view is that if we keep the public informed, if we move forward gradually, then, as and when the benefits accrue, the issues will become progressively less contentious. I am actually extremely optimistic about the future of medical research. The investment of the last fifty years in basic biological research is going to deliver in very big ways in terms of human health and the quality of people's lives.'

In conversation with Norman Levitt

On 18 May 1996, physicist Alan Sokal made the front page of the *New York Times*. The cause of his celebrity is by now well known: he had hoaxed the leading cultural studies journal *Social Text* into publishing a spoof article which parodied the relativism – the idea that things are only true or false relative to particular cultures or discourses – characteristic of much of the work done in university humanities departments. However, what is less well known is that the inspiration for this hoax was a book called *Higher Superstition* written by Paul Gross and Rutgers University mathematician Norman Levitt.

The target of Gross and Levitt's book was the systematic misunderstanding and misrepresentation of science which was typical in the fields of cultural studies, literary theory, science studies, and the like. How, I ask Levitt, did he first become interested in these kinds of issues?

'What first drew my attention to the peculiarities of academic humanists was the de Man affair in the States,' he replies. 'Paul de Man was an eminent literary scholar at Yale University; he was the person who made deconstruction mandatory in English departments. After his death, it emerged that during the Second World War – he was Belgian, originally – he had been a collaborator with the Nazis. He wrote pro-Nazi journalism. But he had never come clean about this fact. Indeed, when he came to the United States, he passed himself off as having been something of a resistance fighter.'

'The fact that his misdeeds had only come out after his death was scandalous enough,' Levitt continues. 'But what happened then was that all his disciples and admirers – all the expected literary characters and also a number of historians and social thinkers – rallied to his defence. Rather than taking a hard look at human folly, which would have been the appropriate response, they plunged into mystification, hypocrisy and obtuseness. Amongst the claims that they made in de Man's defence was one which had it that by writing pro-Nazi, anti-Semitic articles, he was, by some uncanny deconstructive legerdemain, actually denouncing Nazism and anti-Semitism! It was quite absurd.

'I watched this from the other side of the campus as evidence of some very strange goings on. Then my university, along with many other American universities, set up a cultural studies centre. It was decided that its first year would be devoted to science. By this was meant the cultural study of science and the new sociology of science, so we're talking here about people like Bruno Latour and Donna Haraway. A leaflet was circulated, and I think I was probably the only person on the science campus to look at it. I was a bit irritated by it, so I decided I wanted to participate as the local curmudgeon and naysayer.

'In the end, it was determined that I could hang around and perhaps give a talk. Well, the more that I heard what the science studies people had to say, the more I became convinced that there was some very slovenly scholarship going on, mostly devoted to the claim that science was some kind of illusory, ideological construct. I got more and more annoyed by this stuff, and I was toying with the idea of writing a book about it. Then two things set me in motion. The first was the fact that my daughter became seriously ill, with a disease which twenty or thirty years ago would almost certainly have been fatal, but which was now treatable. This really brought it home that there were important matters at stake here. The second thing was that I read some articles by Paul Gross, which dealt with similar themes, so I wrote to him, we started corresponding, and I found that he had had a very similar experience with his own university. We decided that we would have some fun by writing a book, and *Higher Superstition* was the result.'

A common reaction to *Higher Superstition* has been to treat it as some kind of right-wing attack on left-wing thought. But presumably this isn't right?

'Absolutely, it isn't,' Levitt confirms. 'I've been interested in politics for most of my life, but I'm on the Left politically. And neither Paul, nor I, is an ideologue. But I did have a political agenda, it just wasn't the standard one of which we're usually accused. My agenda arose from my distress that the Left had isolated itself and was burying itself in nonsense and circularity, which meant that it was disengaging from any real kind of contact with politics. Identity politics and radical feminism were a part of this story in the States. And I was very upset by it. It was a disengagement from any hope of actually influencing events in the real world.'

This sounds very similar to Alan Sokal's declared intention that he wanted 'to defend the Left from a trendy segment of itself', and his call for it to 'reclaim its Enlightenment roots'.

'Well that's not entirely coincidental,' explains Levitt. 'Sokal had read our book, and he wrote to me saying that he too had suspected that it was part of a concerted attack on the Left, but that he had then checked out the references and had come to the conclusion that things were actually a lot worse than we had said. He saw that campus left-intellectuals were slipping into a kind of delusional megalomania. This was his motivation for writing the parody.'

In *Higher Superstition*, Gross and Levitt carefully document the follies of the academic Left as they are found within the various disciplines of the humanities and social sciences. But has Levitt seen any signs of their influence in his dealings with his own students?

'I think the missionary zeal of the postmodernists has largely backfired, so far as the target student population is concerned,' he answers. 'When I talk to students, I find that there is a certain amount of disgust about the kind of garbage they get from things like freshman composition courses, which have been major venues for postmodern sermonising. But this has eased off in the last few years. However, a little while ago, I was invited to take part in a course that was essentially a morale building exercise singing the praises of multiculturalism. The course supervisors were well-meaning, but when I mentioned to the class that the Egyptians weren't particularly black, there were cries of outrage from the black students. They had been told that the Egyptians were a great black civilisation. This is actually part of the standard line in the Afrocentric world view. The belief is that Europeans had come over via the Mediterranean and stolen everything in Western civilisation from the black world, principally, as they suppose, Egypt. I'm a great admirer of the Egyptians myself, but some of the claims which are made on their behalf are just nonsensical.

'The trouble is that there has been a great deal of indoctrination around this view of history. And a lot of people have believed it. We're talking here about 18- or 19-year-old kids from poor educational backgrounds, and they just weren't used to thinking critically about evidence and argument. I tried to talk to this particular class about the use of steel technology in West Africa, which happened in the third or fourth century BC, and which was actually very impressive. But the black kids wouldn't listen. Once I'd said that the Egyptians had very little to do with black West Africa, they were horrified and outraged, and that was that.'

Many of the criticisms of science voiced by those on the academic Left are predicated on an argument which reasons something like this: science is inextricably a social process, with all that this involves in terms of power relations, biases, prejudices, and so on. Its truth claims, therefore, are necessarily partial and relative only to its particular discourses. What, I ask Levitt, does he make of this kind of argument?

'Well, obviously, there is a certain amount of common sense about it,' he replies. 'Not everything in science is about disinterested evidence and dispassionate argument. There is a lot of dirty work which goes on, a lot of yelling and screaming. In mathematics, for example, you see great resentments and controversies emerging, usually over priority. Not every controversy is easy to settle. Some linger for years.'

'But amazingly,' he continues, 'a lot of them *are* pretty easy to settle. People will abandon a position fairly quickly when the evidence mounts up against it. Occasionally, you do get something like Fred Hoyle's commitment to the steady-state cosmological theory, where he was never really able to accept that the evidence was stacked up against it. But this is unusual. People are not normally ineluctably committed to their theories.'

But what about the established facts of science – for example, that hot air rises – presumably constructivist theories just aren't plausible in these kinds of cases?

'No, really they are not,' Levitt confirms. 'There are times and places where things are unclear enough that arm twisting and throwing your weight around will hold the field for a bit. But after a while, the truth will out. A key part of the scientific ethos is that the young guy will always want to outdo the old guy, and sooner or later – more often than not, sooner – better theories win through. What usually happens then is that the old guy will say something like, "Well, that's what I meant all along!" '

For many people, what defines science is that it deals in theories and hypotheses which are testable, and it is how they match up against the evidence which determines whether they will survive or not. Is that part of what makes science different from the kinds of work which goes on in the humanities?

'Yes, that is a big part of it,' Levitt agrees. 'But there are certain areas where this isn't the case. For example, just about every physicist will have a different idea about quantum mechanics, because it is almost impossible to decide between theories by empirical experiment. This is a great frustration. Physicists who are diametrically opposed on the question of what is going on at the fundamental level will be in complete agreement about experimental outcomes.

'However, in this situation, people are really wearing two hats. There is no disagreement about the hard physics. But the philosophical implications of the physics are up for grabs, at least for now. This is actually very interesting. There are certain questions which are not addressable by empirical means; there are too many possibilities which are consistent with what the data reveals. In this situation, issues of temperament or philosophical inclination do play a part in how the data is interpreted.'

'But, it is important to understand that these are very special situations,' Levitt emphasises. 'Indeed, whilst these are interesting questions, if we never understand, for example, how relativity coexists with quantum mechanics, it won't really affect anything. It doesn't have material consequences. That's part of the problem: you can't devise an experiment to see what's going on. This is very frustrating for people who care about these things, but a lot of physicists, and certainly a lot of people up the line in chemistry or biology, really don't care, and they tend to view these arguments with ironic disdain.'

As well as the more general claim that scientific truths are necessarily culturally embedded, the academic Left make a lot of the relationship between capitalism, big business and science. There is the suspicion here that they might be onto something. For example, in the issue of *The Lancet* dated 8 April 2000, researchers from the University of California at San Francisco reported that there was evidence that the tobacco industry had closely monitored and had tried to interfere with the conduct of a study on lung cancer in non-smokers. Obviously, this doesn't suggest any radically relativistic conclusions, but it does seem plausible that power and politics will sometimes play a role in determining which theories particular scientists are likely to embrace. If a study is being sponsored, for example, by a tobacco company, isn't it then more likely that its findings are going to be to the liking of the company?

'I'm afraid this is a function of the way that institutional science works, and it is worrying,' agrees Levitt. 'In the United States, there is enormous commercial pressure, and where there are working agreements between private corporations and university researchers, things can get badly skewed. For example, there has been a lot of corporate machination to shut researchers up using little known by-ways of American law. And this is a real danger. Among the accomplices are American universities, whose eagerness to turn a profitable deal often overrides good sense and scholarly conscience.

'But if this were what the academic Left were talking about, then I would never have been particularly antagonised by them,' says Levitt. 'However, it was not what they were talking about. Other people talk about it on the Left, and there is also a lot of concern about it in the scientific community, but it wasn't really the agenda of the post-structuralists, the social constructivists, the feminist epistemologists, and so forth.'

Is there a solution then to this problem?

'There is no easy solution, but it's got to start with some of the more prosperous universities taking a stand against it,' he replies. 'They can afford to do this. The problem is with the middling universities. For them, the lure of a profitable deal is more tempting, and it does result in things getting distorted – not so much in terms of scientific theory, but rather in the context of who is hired, what they are set to do, whether research is of scientific, as opposed to commercial, importance, and whether results are published. There are notorious cases where people have been forbidden to publish results because they went the wrong way in terms of their likely effect on share prices. This *is* something which has engaged the attention of various scientific societies. It's a fight, but it is a necessary fight.'

What is the attitude of journals towards this? Are they aware that there are these kinds of forces at work?

'They've tightened things up a little bit,' replies Levitt. 'Especially after some of the scandals which have erupted. People are aware of the importance of refereeing standards, the high levels of scrutiny required when commercial ventures have something on the line, and the perennial problem of the multi-authored paper. The system in biomedicine, for example, is positively feudal. It has been the locus of the worst kinds of abuses, where people have had to jump through hoops in order to keep the heads of laboratories happy. Mathematics is at the other end of the spectrum; great indulgence is given to 20-year-old hotshots, and older people will often volunteer to take on teaching loads, in order to free up the time of younger people so that they can get their research started.'

One of the central themes which runs through *Higher Superstition* is that the level of scientific knowledge amongst people working in the humanities is pretty poor. If this is right, it mirrors a general lack of scientific literacy amongst the population as a whole. For example, a

National Science Foundation (US) survey conducted in 2001 found that 25 per cent of respondents thought that the Sun went around the Earth rather than the other way around. Does this kind of thing worry Levitt?

'I did come to take the view that what one was seeing in academic life was a reflection, in its own particular colours, of a more general social phenomenon,' he responds. 'People have the sense that science is very important – and it's obvious why they might think this – but there is a frustration with the difficulties of learning scientific particulars. I think this is partly why the academic subculture of science studies – pretend expertise, if you like – has sprung up.

'In lots of ways, this occurs in popular culture and politics as well. People try to take some control, at least subjectively, over something which they know is important, and which impinges on them, but over which they don't have any real mastery at a certain conceptual level. So you see a lot of popular movements asserting this and that on technological or health questions, borrowing scientific jargon and then pulping it into a pseudoscience. And people who should presumably know better sometimes go along with this to a shameful degree. In the States, for example, there are lots of perfectly respectable hospitals with alternative medicine departments.'

Levitt mentions that the protests over genetically modified food which have occurred in Europe are indicative of this phenomenon. 'It is very interesting. Why should this have promoted so much concern? Many things run through it. Mary Douglas, in her book *Purity and Danger*, puts forward the idea that cultures assume there are "natural" categories, the transgression of which will bring about retribution. This obviously underlies much of the uneasiness concerning GM foods.

'The argument in favour of GM technology is obvious. The people who are loudest in their complaints about it are the people who are most wrong about it. From a hard-headed, ecological point of view, the pressure on the environment comes because subsistence farmers cannot make a living. They continually have to move to find land which will support crop cultivation for a couple of years. That's what's killing the rainforests. If you're concerned about the preservation of wilderness areas and their biota, then you want to find cultivation methods which maximally exploit what is available, and in poorer countries you're not going to do this with heavy machinery; you have to use genetically modified cultivars.'

Is it, therefore, desirable that scientists should address this lack of scientific literacy in the population? Would we get more rational thinking about things like GM crops, if they did?

'That's a tough question,' Levitt admits. 'How much time and energy have people got for this kind of thing? Also, in a sense, it is true that a little learning is a dangerous thing. The most virulent activism is often led by people who know enough to speak a little of the language of science. So there is no guarantee that increasing what people know will lead to more rational thinking about these issues.

'I'm actually very pessimistic about the whole situation,' he says. 'A couple of years ago, I addressed a conference of US federal judges on what they should know about science in order to make informed decisions on various matters where science is involved. They were very bright people, obviously. Very hardworking. But basically, they were pig-ignorant about science. For example, they had no idea about what DNA evidence could show.

'I was there to talk about mathematics and statistics. It was clear that the learned judges were susceptible to the same kind of fallacies as the average man in the street. People are very poor at inference from quantitative evidence. For example, they see a cluster of leukaemia deaths, and they assume there must be some causal factor at play. If you know something about statistical inference, then you know that the odds are that there is no such factor; that it is just a matter of chance. But I saw these really rather bright jurists having a great deal of difficulty with basic probabilistic reasoning.'

Levitt's pessimism here dovetails in an interesting way with his views about the ability of non-scientists to understand enough about science to enable them to write about it properly. The suggestion in *Higher Superstition* is that people cannot write a successful social history of science unless they have achieved near professional competence in the science they're writing about.

'Professional competence means that you understand what the hell people are talking about,' Levitt says. 'This is really what is at issue. There are some areas – the history of technology, for instance – where things aren't quite so clear cut. But if, for example, you're trying to write a history of Newton's disputes with Leibniz, then if you don't know calculus, you're in trouble. You can't see precisely what's at stake, and what might count as evidence for one thing or another. If you're not competent in the field that you're writing about, then you're engaged in trying to understand the thoughts of people whose mental processes are completely alien. Of course, this doesn't mean that you're automatically a good historian if you're a mathematician. But if you have a solid background in mathematics, as well as the other relevant scholarly skills, then you're in a strong position if you're writing a biography of another mathematician.'

What then does Levitt think of the popular science writing which is produced by non-scientists?

'A good deal of the popular literature is poorly done,' he replies. 'But there's been a trend, nurtured by people like John Brockman in the States, to bring in prominent scientists – Richard Dawkins and Steven Weinberg, for example – to write their own books. This is the better way to go, though you have to watch out for people blowing their own horn.'

No doubt it is significant that the vast majority of 'science scepticism' comes from within the humanities and social sciences, with very few scientists joining in. However, this is not exclusively the case. Particularly, for example, the debates over sociobiology have led to radical disagreements amongst scientists about the scope of scientific explanation, and also about the part which politics plays in scientific theorising. Is there, I ask Levitt, something about sociobiology which makes it susceptible to these kinds of more radical disagreements?

'It is certainly true that these guys are politically motivated', he replies. 'Richard Lewontin is probably the last living dialectical materialist! But these arguments are actually the continuation of a long debate. About fifteen years ago, an intellectual historian, Carl Degler, wrote a book called *In Search of Human Nature*, tracing the vicissitudes of theories of innateness and in-bred human propensities, describing how they had fared over the years depending on political circumstances and fashion. This has been going on for some time. Part of the problem is that within sociobiology there has been a lot of highly visible popularisation, with great claims made on the basis of fairly ambiguous evidence. And another part of the problem is that people like Lewontin, and various feminist theorists, just don't seem to want to hear what sociobiology has to say. It undercuts too many utopian daydreams.

'My view is that sociobiology is a good programme, but it has got its work cut out for it. It needs to establish a methodology which will be generally accepted by its advocates as producing solid evidence. And it needs to avoid making great pronouncements about social and political issues on the basis of tentative theories. I think we'd get much further with it, if partisans on all sides of the issue were silent, and let those who want to work on the stuff get on with it in a quiet, scholarly way.'

One of the criticisms that people like Steven Rose and Richard Lewontin, on the science side, and Hilary Rose and Mary Midgley, on the humanities side, make of sociobiology is that it is inappropriately totalising in its scope. Particularly, it is argued that sociobiology fails to recognise that there are aspects of human behaviour which are best examined with techniques other than those of the empirical sciences. What's Levitt's view on this matter? Is the phenomenon of 'being in love', for example, susceptible to a reductionist, mechanical explanation?

'Well, one could certainly believe that love is ultimately just a matter of neurons firing in response to stimuli, and so on; and in some broad sense I think that this is the case,' he replies. 'I am a monist, I think there is only one kind of reality. But, of course, we don't yet have anything like this kind of explanation of love.

'Also, the whole business of reductionism is often poorly understood. It is possible to do perfectly good science without being strongly reductionist. But this is a different issue from one's *ontological* commitments. I'm surely not urging every researcher in economics to study everything at a neuronal level before they tackle the psychology of the market; that would be silly. But I *am* suggesting that the world follows a certain causal chain, and that taking this into account will lead to important scientific developments through stronger theories and deeper explanations. This isn't the same as saying that *every* explanation has to be reducible to physics. Reductionism is an opportunistic strategy. When it's available you use it, but there are lots of other techniques to use as well.'

The suggestion in *Higher Superstition* is that the machinations of the academic Left are mainly a local worry concerning academic life. However, it is possible that this is slightly underplaying what's at stake. For example, the science writer Meera Nanda has pointed out that by advocating the validity of 'local ways of knowing', the academic Left have given moral support to right-wing, Hindu fundamentalism, with all that this means in terms of the caste system, the subordination of women, and so on. Does Levitt recognise that there are these wider issues?

'Certainly, there is a bit of this,' he replies. 'Meera is worried about what is happening to Indian social movements – movements she has been involved with – where a lot of the struggle has been against traditional views, and where modernisation has been a great ally of social progress. What she found was that a lot of left intellectuals in India have recently embraced a relativist view of the world, exalting local practice and knowledge – which in India means a fairly vile social system. I'm not certain whether things would be very different had the Left not been influenced by relativism – the political and social problems go deep – but certainly it didn't help that the small cadre of intellectuals had become imitation deconstructionists and Foucauldians.'

'In fact,' he continues, 'we have had the same kind of thing going on in the United States. Left intellectuals have claimed that native Americans should maintain their native beliefs and practices, and abjure all mainstream American culture, thereby preserving a beacon of alternative human possibility. But, of course, the people that you're asking to do this are not living happy lives. They are in many cases living nasty, short and brutish lives by modern standards.'

Since Gross and Levitt wrote *Higher Superstition*, a new threat to the integrity of science has emerged in the United States. The 'intelligent design movement', organised around The Discovery Institute in Seattle, seeks to replace Darwinian explanations of the evolution of species with the idea that species are designed by an intelligent agent. Needless to say, if the intelligent design people are pressed on the matter, it turns out that this agent is the God of the Christian Bible. Is this something which worries Levitt?

'Yes, very much so,' he answers. 'The ID movement is very clever politically. They've gathered together a few people who supposedly have scientific credentials – unfortunately, many of them are nominally mathematicians – and they claim to have wonderful arguments which show that Darwinian theories don't work. Of course, it's nonsense, but they're talking, for example, to congressmen, who may already believe in creationism, or who, for political reasons, may want to go along with it. And suddenly, they've got something which is saleable. It is very dangerous.

'Indeed, it is much more worrying than the academic Left ever were,' Levitt says. 'The alarming thing about the academic Left wasn't that they were going to castrate science directly, but simply that they were stripping away the institutional support that science requires. But these ID bastards really do want to castrate science. Interestingly, the academic Left have been scared by the rise of the creationists. They're smart enough to see that they're dealing with some pretty nasty animals. And, as a result, they have lost a lot of their anti-science zeal. On the other hand, ID propagandists have made clever use of the postmodern dictum that scientific fact is merely in-group social consensus. The story they tell is that a cabal of godless materialistic scientists has connived to deny the evidence for the mind and hand of God. The irony is that the Left have done the spadework to prepare the ground for this right-wing mischief.'

11   Biodiversity

In conversation with Edward O. Wilson

The twentieth century saw huge advances in our knowledge and understanding of the world. For example, special and general relativity, quantum theory, antibiotics, the structure of DNA and the Big Bang theory are all accomplishments of this period. Yet despite this progress, there remains, of course, a lot about which we are in the dark. In some areas, we lack knowledge and understanding for foundational reasons. For example, physicists currently have little idea about how to get real-life, testable propositions out of superstring theory (currently the best candidate as a *theory of everything*). In other areas, our ignorance has much more to do with the extent of the scope of our *possible* knowledge, rather than any specific barriers to understanding.

Biodiversity is a case in point. 'One of the most remarkable facts about life on Earth is just how little we've explored of it,' explains Edward O. Wilson, Professor of Science and Curator of Entomology at the Museum of Comparative Zoology at Harvard. 'The number of species which we have described and named is thought to lie between 1.5 and 1.8 million; but even here, no complete census has been made on the basis of all the available literature. The *actual* number of species is not known to the nearest order of magnitude. Various estimates have been put forward by different specialists, ranging from an improbably low 3.6 million to an impressively high 100 million species. The fact that one recent estimate has it that there are approximately 4 million species of bacteria, almost all unknown, in a single ton of dirt, suggests that the higher figure is probably nearer the mark.'

If then there are close to 100 million different species, how many individual organisms does this add up to, or is it impossible to specify?

'I think the term "astronomical" would not be inappropriate here,' Wilson replies. 'One estimate – a few years old now, but based on rigorous research – suggested that there are approximately 10^{18} insects alive at any one time. One could add to that an order of magnitude estimate of the number of microorganisms; we know that there are something like 10 billion bacteria in a single gram of soil. And then we'd have to include the bacteria and microscopic flora which we've recently discovered live down to a depth of at least two miles below the surface of the Earth. So "astronomical" is the right word.

'It is clear that we live on a little known world,' Wilson continues, 'which means that a large part of biology in the twenty-first century must be about exploring this planet as quickly as possible with the aim of completing the Linnaean enterprise. Another way I like to put this is to note that something like 10 per cent – though nobody knows for certain – of species have been described in the 250 years since Linnaeus established his classificatory system, which means that some 90 per cent remain yet to be described. How many of the 90 per cent are small invertebrates, crucial in the maintenance of the ecosystem, and how many are microorganisms, which are extremely important for human health and other aspects of human welfare? Well, we just don't know, which means that exploring planet Earth should be a priority issue in the twenty-first century.'

Although the concept of the 'species' is fundamental to the Linnaean binomial system (the system by which organisms are classified), it is, in fact, something of a contested concept.

'How to define a species has troubled biology right from the start,' Wilson tells me. 'At the present time, for organisms which reproduce sexually, we use what is called the *biological species concept*. In simple terms, a species is a population, or series of populations, which is reproductively isolated, thus there is no interbreeding with other populations under natural conditions. This works for a wide range of eukaryotic organisms – the ones most familiar to us – so the model has served us very well.

'But there are many organisms for which the biological species concept does not work so well. For example, there are many species which are asexual. In these cases, which include a large number of microorganisms, we employ an arbitrary cut-off point in terms of degree of similarity. For example, a standard which is used with bacteria decrees that if two strains have fewer than 70 per cent of their base pairs in common, then they are separate species.

'However, I don't agree with many of the critics of the biological species concept, who say that it is sufficiently flawed so as to make studying life on Earth very difficult. I have exactly the reverse view. I don't think we should worry too much about forcing every organism, aprioristically, into a settled, procrustean definition. Rather, we should select the definition which seems to be working best now, get on with the job, going along by trial and error, conference discussion, and so on, and then as we come to know more about biodiversity we will arrive at a solution to the species problem. In other words, we should look at the species definition, and the problems associated with it, not as a barrier to research, but rather as an interesting problem in biology which we will in time solve.'

If there are currently some 100 million extant species, what is the history of their emergence? Presumably, notwithstanding periodic mass extinctions, there has been a gradual and general increase in biodiversity over the last three and a half billion years?

'Yes,' replies Wilson, 'and particularly during the last half-billion years, which has seen the emergence of a large number of multicellular organisms, and, during its early phases, the invasion of the land by plants and animals. The increase in biodiversity has been very gradual, though, and there have been many setbacks, including five major extinction spasms. And, significantly, we reached the highest level of biodiversity just before the arrival of humanity.

'There is also an interesting point here about whether evolution involves progress,' Wilson says. 'Some philosophically inclined biologists have suggested that there is no such thing as progress in evolution. But I've never been able to understand why anybody, whether scientist or philosopher, would want to argue about the concept of progress. If progress is defined as being goal-directed, then there hasn't been progress, because evolution is essentially mindless. But if we're talking about progress in the sense of increasing complexity, abundance and diversity – things which most humans would equate with progress – then it has occurred. In this sense, evolution has had a fabulous history of progressive change, moving from the earliest prokaryotes, through the first eukaryotic organisms, to multicellular, to large multicellular, to complex multicellular organisms, and so on, all the way up to intelligent organisms.'

Has all this happened for a priori reasons? Is it built into the nature of Darwinian natural selection that there will be this change from simple and few organisms to complex and many organisms?

'No, not at all,' Wilson answers. 'It could have been different. It could have taken any number of trajectories. Given different conditions, for example, evolution might have resulted in a certain level of complexity, and then stasis. Or there might have been oscillating levels of biodiversity, driven by regular environmental events of a particular kind. So it is possible to imagine a large number of scenarios.'

But if one were modelling evolution, would it be possible to set up the initial conditions in such a way that one would necessarily get the kind of progress which Wilson is talking about?

'The short answer is yes,' he replies. 'The questions that we're addressing here – the modelling of planets, the likelihood of evolution of different kinds – are going to become major concerns of exobiology. We are as close as ten years away from discovering the first Earth-like planets outside our own Solar System. And at that point we'll be doing this kind of modelling work.'

For people who wish to question the veracity of Darwinism, the emergence of new species – speciation, as it is called – is a particular target. Critics claim – erroneously, in fact – that there have been no observed instances of speciation, and that therefore Darwinist explanations of biodiversity are without scientific support. But Wilson is adamant that these kinds of objections amount to nothing.

'We know exactly how most speciation occurs,' he insists. 'We know the dates of the separation of species; we know the varying degrees of the divergence of populations; we know how populations interact; and so on. Denying speciation is like denying the evolution of star systems simply by ignoring everything which astronomy tells us – just so that it is possible to say that God put all those bright lights up there, like we've always been told.

'The misinterpretation of the accumulated knowledge of biology doesn't deserve much attention,' Wilson says. 'I've read Michael Behe's *Darwin's Black Box*, where he advances the idea of irreducible complexity. What this amounts to is that Behe can't imagine how certain steps in evolution might have been taken, and this is supposed to be construed – although he would deny it – as a powerful argument for a loving and personal Judaeo-Christian God. It is absurd.'

The trouble is, though, that many people outside the scientific community in the United States take the claims of people like Behe – and the intelligent design community, more generally – quite seriously. There was, for example, the notorious decision of the Kansas Board of Education, now reversed, to remove evolution from the state's science curriculum. Does this kind of thing worry Wilson?

'It worries everybody,' he replies. 'Evolution is a fact. It occurs. It should be undeniable. There are innumerable lines of evidence, so to refer to evolution as being non-scientific is just false. I was raised a creationist, and I have the following response to people who doubt evolution. You have to grant, if you've got any intellectual honesty, that there is massive evidence for Darwinism; that it is irrefutable that the evidence suggests that the complexity of life has been created by evolution. I have read the Bible, and I have listened to many sermons, and I know that the God of sacred scripture has often been paternalistic, often angry, often importunate, often warm and supportive, but He has never been tricky. I cannot believe that He salted the Earth so thoroughly with false cues, in the knowledge that it would mislead us, and that those people who came to support evolution would be considered impious.'

Ironically enough, the very biodiversity which is the product of evolution, and the source of the evidence for it, is being placed under some threat by the activities of human beings.

'The first thing that we did, as we started to come out of Africa about 50,000 years ago, was to kill off the mega-fauna,' Wilson explains. 'And over time we have had this kind of impact on increasing numbers of animals. In the last few centuries, we have developed the ability to transform habitats, including the rainforests, into agricultural land; and having done so, we tend to overuse the land. We are also beginning to wipe out the smaller organisms, including many insects, and also plants.'

'All this is very much against our own self-interest,' he says. 'We should bear in mind, for example, that insects are central to our existence. Insects are almost exclusively terrestrial creatures; they live on the land. They are the primary turners and enrichers of soil; the primary scavengers of small, dead animals; the primary pollinators of plants all around the world, including our own crops. If you take away insects, then the vast majority of plant systems will disappear. In an essay called "The Little Things that Run the World", I have argued that if insects were magically removed from the Earth, then the living world on the land would collapse within a year or two, and with it humanity. Humanity would not survive the extinction of insects.'

Is it possible to find evidence that this is a worry which we need to take seriously now, rather than its simply being something which we have to bear in mind for the future?

'Well, it is certainly possible to notice the effects of a reduction in the diversity of insects in some parts of the world,' Wilson replies. 'For example, we have discovered that when honey bees are hit by disease or parasites, and this has happened in areas of the United States, crop productivity can be drastically reduced. We also know that when crops are treated too heavily with pesticides, killing insects, spiders and other creatures, there can be outbreaks of serious insect pests; while there are a smaller total number of species, each has more opportunities open to it, which means that you get a larger number of crop destroyers and disease vectors. The signs then are that if we permanently reduce insect diversity beyond a certain point, there will be catastrophic effects on the parts of the environment we depend on.'

A response which is sometimes offered to people who raise these kinds of concerns is to claim that there is no need to worry too much about the environment, because the nature of human progress is such that we will always find technological solutions to problems which arise. Does Wilson have any sympathy with this kind of argument?

'Absolutely not,' he answers. 'I have an expression for this, which is: "One planet. One experiment." It is equivalent to gambling with just one throw of the dice, where everything is lost if they come up snake-eyes. We shouldn't be throwing dice at all; we should be looking at the world's environment much more closely to find out how to keep the planet safe. We can be completely anthropocentric about this; we need to pay close attention to the living environment and biodiversity for our own good.'

One of the interesting facets of biodiversity is the geographical spread of species. The evidence suggests that the majority of species of animals and plants live in tropical forests, especially tropical rainforests. However, as a result of deforestation, tropical forests are disappearing at a rate of 1 to 2 per cent a year. In his book *The Diversity of Life*, Wilson worked out, on the basis of conservative parameters, that approximately 0.25 per cent of the species in rainforests are condemned to early extinction each year. Deforestation then is clearly a major threat to biodiversity. What, I ask, are the human forces driving deforestation?

'Ignorance, greed and denial, in that order,' Wilson replies. 'Most deforestation is a result of the actions of local and corporate people who don't know what they're doing. However, many multinationals and other corporations know exactly what they're doing, but to make money they just plough on ahead with deforestation anyway. For example, there are logging companies, "rogue" or "maverick" companies, which deliberately steal away the timber from poor developing countries by the mechanism of cheap logging leases. And then there is denial; there are still a lot of people who say that the bad effects of deforestation are not proved, or that they need more information. That's the line of the Bush administration, for example.

'The irony of deforestation is that it has been well documented that with just a little bit of planning and market development most tropical rainforests can be made more profitable by sustainable maintenance than they can by logging. For example, the latest figures we have available from the National Forest of the United States show that it contributes 35 billion dollars to the gross domestic product. Of that, 78 per cent came from recreation, including hunting and fishing, and only 13 per cent came from logging. More and more developing countries are discovering that with just a little bit of cleverness and planning, and a little bit of political stability, they can start to make a lot of money from tourism alone.'

So in terms of an ethics of conservation, the choice isn't between conserving and being less well off or not conserving and being better off, but rather it is possible to conserve and to do well economically?

'That's right,' Wilson confirms. 'Also, in most developing countries, given the present state of the world, it is not going to be possible to make an effective, purely ethical argument about conservation. This might change eventually, but right now it is much better to present the ethical argument with an economic bottom line attached.'

This ties in with something which Wilson says in *The Diversity of Life*. He notes that the movements associated with new environmental thinking are realistic about the impact of things like poverty and population growth on the potential for proper conservation; and that it is generally understood that it is necessary to develop fairly concrete proposals to show how it is possible to meet the aspirations of a world populated by up to 9 billion people without destroying vast tracts of natural habitat.

There is an interesting point here about the attitude which environmental groups such as Friends of the Earth have towards genetically modified crops. To the extent that GM crops result in increased yields, they offer the potential of alleviating some of the problems associated with hunger, and so on, yet most of the various conservation movements are vehemently opposed to GM technology. Does Wilson think that there is some muddled thinking going on here?

'Yes, I think the opponents of genetically modified organisms (GMOs) have got it wrong,' he says. 'Subsistence farming puts pressure on natural environments. The evidence is overwhelming that in order to save the few remaining and rapidly shrinking natural environments we need higher crop productivity and market development in the agricultural systems of developing countries. We have a lot of natural species which, if they can be made marketable, can be used to start this process. But we can also take existing crops, the twenty or so leading species, and by means of genetic modification prepare strains of them which are variously perennial, nitrogen-fixing, saline-resistant, suitable for dry farming, and so on. This is a strong, positive reason for GM crops. They will help the economies of developing countries, and they will take some of the pressure off their natural environments.'

'It is, of course, necessary to pay careful attention to the negatives associated with GMOs, and the situation must be carefully monitored,' Wilson says. 'But my view is that the pluses greatly outweigh the minuses. I'm inclined to think that a lot of the reaction against GMOs has been because people focus too much on the possible risks and don't weigh them up properly against the potential advantages.'

Is there also perhaps a sense in which environmentalism for some people is a quasi-religion, and that you, therefore, get a reflex anti-GMO position?

'I think there is an element of this,' replies Wilson. 'But I've got to say that I've had very little opposition from my fellow conservationists and ecologists. I think that is perhaps because they know that my heart is pure. In other words, they don't see me as someone who is arguing for a particular faction or interest group. I tend to think that although there are extremists among environmentalists, most of the key people, even the activists, want results, and they will be inclined to weigh these things up quite carefully.'

In his book *The Future of Life*, Wilson outlines some of the key elements of a solution to the problem of the destruction of natural habitats. Amongst other things, he talks about the need: to salvage those 'hotspots' which are at greatest risk of being destroyed; to keep the five remaining frontier forests intact; to stop logging in the old-growth forests; to make conservation profitable; and to use biodiversity to benefit the world economy. How achievable are these things?

'In the last twenty years, we've seen a remarkable and heartening growth of global conservation organisations,' he answers. 'And there has also been good progress in the technologies for studying the state of the environment, including biodiversity, and in new methods, some still experimental, for instituting conservation techniques. It turns out that effective conservation has to be conducted through what is an "iron triangle", comprising science and technology, the private sector and the government. If you can get these groups aligned, then you will get conservation.

'Our experience tells us that when we've done proper research, and we know where the worst problems are in terms of habitat destruction and species elimination, then we can bring governments on board – especially in small countries – through economic advice, and through the exchange of logging concessions for more profitable conservation concessions. If we can add to this an increasing awareness of the importance of conservation on the part of the nationals of a country, then we can get rapid progress.

'However, this work is at an early stage,' Wilson cautions, 'and we have been deliberately picking easy targets, like Surinam and New Guinea. Each major environment has to be treated separately and individually, and approached in its own optimum way. There are going to be some very tough problems ahead. As it stands, we're not even 10 per cent of the way towards where we should be in terms of global conservation.'

There is the perception that the Bush administration in the United States, with its neoconservative agenda, has very little interest in either conservation or environmental issues more generally. Is this something which worries conservationists?

'Yes, it is a worry,' admits Wilson. 'It is their inaction and ignorance which is the worry, as much as specific opposition to the things which we've been saying. In terms of American domestic and foreign policy, the environment has been taken off the agenda. It is not just a neglect of our own forest reserves, national parks and biodiversity reserves, but also a neglect in our foreign policy, including our foreign aid policy; our market development and trade policy; and in the question of how best to help and influence countries, as opposed simply to policing them. The neglect of all of these major policy areas with regards to the environment means lost opportunities of considerable magnitude, at what is undoubtedly a large cost to the future.'

What underlies this apparent lack of interest in environmental issues? Is it an ideological thing connected to their neoconservatism, perhaps something to do with their commitment to minimal government?

'I'm not the best person to ask for a political analysis, but I think it is largely ideological,' Wilson replies. 'There is, of course, a certain amount of narrow self-interest at work here: paying off corporate supporters, that kind of thing. But mainly it seems to be to do with an ideological commitment to pragmatism, minimal government and a developing anti-science position.

'The pragmatism aspect involves them saying something like: "We are problem solvers. Admittedly, these are short-term solutions. But we think that the best way to move this tired old Earth along, and to bring peace, is to solve the problems which are before us right now. We don't see the environment as being a priority. You might be right that there will be problems down the road, but we've got time on our hands. The major issues now are terrorism, preserving American markets, and things like this." So that's their pragmatic side.

'Minimal government is part of the philosophy of the current brand of right-wing conservatism. Their claim is that there is too much government in the United States. They argue that it is necessary to cut down on bureaucracy and regulation, in order to return power to the people. I won't comment on this disastrous way of thinking!

'Their anti-science, for its part, has a strong religious aspect to it. There is the feeling that the United States should return to old-fashioned family and religious values, and that somehow scientific materialism is responsible for their erosion. So they've decided to make everything a hard-sell for scientists. Bush's response when warned about environmental issues was something like: "Well, this is what the bureaucrats say, but we're not sure about it."

'I think that all these factors are at work in their marginalisation of conservation and the environment.'

What is Wilson's best guess about the outcome of all this? What course of action will humanity choose with regards to biodiversity and conservation? Will we come through the problems that we're currently facing with the natural world largely intact; or will people look back at us from the standpoint of the twenty-second century and accuse us of destroying the world's evolutionary heritage?

'Obviously, this is at bottom a political and sociological issue,' Wilson replies. 'It comes down to how we respond to the problems which confront us. I'm cautiously optimistic. The reason I'm optimistic is that we've got the science, we know the problems, we know how to solve them, and we know how much it is going to cost. It has been estimated that to save 70 per cent of the animal and plant species of the world, including the hotspots which are most under threat, will cost roughly 30 billion dollars, one part in a thousand of the world's gross annual product. You would think that it would be considered a godsend that it is possible to save a large part of life for what is, relatively speaking, a minute sum of money. So this is the reason for guarded optimism. But it has to be guarded, because the world is turning out to be just much more irrational than any Harvard professor would want to believe.'

In conversation with Martin Rees

The cover of the June 1947 issue of the *Bulletin of the Atomic Scientists* featured a clock showing the time seven minutes to midnight. It was, in the words of an editorial of the *Bulletin*, supposed to represent

> the state of mind of those whose closeness to the development of atomic energy does not permit them to forget that their lives and those of their children, the security of their country and the survival of civilisation, all hang in the balance as long as the spectre of atomic war has not been exorcised.

During the decades of the Cold War, The Doomsday Clock, as it became known, came to symbolise the follies and dangers of an arms race which resulted in the world's two great superpowers stockpiling enough nuclear weapons to destroy the Earth many times over. As it has turned out, nuclear disaster has so far been averted. But, as Martin Rees, Astronomer Royal, and Professor of Cosmology and Astrophysics and Master of Trinity College, Cambridge, indicates in his book *Our Final Century*, this has probably been more a matter of luck than judgement.

The thought that humanity was lucky to emerge from the twentieth century without having blown itself to pieces raises a whole series of questions about the role of science in society. Particularly, it brings to the fore the issue of whether scientists are responsible for the uses to which their discoveries are put. The atomic scientists were obviously well aware of the ethical and social dimensions of their research. For example, in 1957, a group of them, from many different countries, attended the first of the Pugwash conferences – still running to this day – which have the aim of looking for cooperative solutions to world problems.

'I think that group of scientists behaved very well,' Rees tells me. 'They felt that it was their responsibility to engage in dialogue with politicians and the public to ensure that the risk of nuclear war was minimised.

'My view is that their example should be followed by scientists in other fields, particularly if they're working on things which might in the future have as epochal an impact on civilisation as nuclear weapons did. Scientists can't control how their work is applied – and nor should they, because the decisions on the use of science should be made by a wide public – but they have a duty to engage with the public in pointing out the benefits and risks of what they're doing, and in urging benign uses of their discoveries. The progeny of a scientist are their ideas and they ought to be concerned about the fate of these ideas, even if they can't fully control what happens to them.'

This view that scientists have a responsibility to keep the wider public informed about their work, and to urge benign uses of their discoveries, ties in with the major claim of Rees's *Our Final Century*, which is that there is a distinct possibility that humanity will not make it out of the twenty-first century. What, I ask Rees, was his motivation for writing this book? Was it related in any way to his work in cosmology?

'Well, for the last twenty years or so, in parallel with my research, I have been writing articles and giving talks on general issues like the responsibilities of scientists, and the opportunities and risks brought by science,' he replies. '*Our Final Century* is really a culmination of this process.

'Nevertheless, I do think the fact that I'm a cosmologist has given me one special perspective: namely, a sense that we might be at a very early stage in the emergence of complexity in the cosmos. Although most educated people will be aware of the enormous span of time in the past, and how, for example, it was necessary for the emergence of the Earth, and for Darwinian evolution, they tend to think that humans are somehow the end point of this process. However, wearing my cosmologist's hat, I am very aware that even our own Earth may have a future which is longer than all the time which has passed to now, and that humans might be a very early step in the development of complexity and consciousness in the Universe.

'Also, it's an important point that what is happening here and now on Earth, even in the cosmic perspective, is very special. If you had been watching the Earth over the four and a half billion years of its existence, then for the vast majority of this period you would have been watching very slow and gradual change. However, over the last 100 years, there have been dramatic changes: increases in the levels of carbon dioxide, the emission of radio signals from Earth, and, for the very first time, small lumps of metal escaping from the Earth's atmosphere to go into orbit around the planet and into space. There is then something very special happening on Earth, a kind of spasm occurring less than half the way through the planet's potential life.

'If you were then to think about what might happen in the next century, there are a number of plausible scenarios. It is possible that humanity might move to a situation which was sustainable into the future; or we might see the first steps in expanding the biosphere, including human consciousness, beyond the Earth; but it is also possible that humanity will visit destruction upon itself, and that there will be a regress in civilisation.'

Part of what makes the stakes so high as far as the future of humanity is concerned is that it is not yet known whether the kind of complexity which brought about intelligent life on Earth exists anywhere else in the Universe. Rees has argued that it is sensible to be agnostic about this question. What led him to that conclusion?

'Mainly it is a matter of the limitations of our present knowledge,' he answers. 'All living organisms have layer upon layer of intricate structure, far more complexity than anything which exists in the inorganic world. Although we understand how natural selection led from simple organisms to our present biosphere, there is no general consensus about how life got started in the first place. We don't know whether it was inevitable or whether it involved some large fluke. Therefore, we cannot say how likely it is that life would get started in another environment. Also, even if we begin at the point at which life gets started, we don't know whether it is inevitable that intelligence will eventually emerge. So whilst progress in astronomy in recent years means that we now know that there are large numbers of planets like the Earth orbiting other stars, we still don't know how likely it is that life would get started on any of these or how it would evolve if it did. For these reasons, then, I think agnosticism about the possibility of life elsewhere in the Universe is the most sensible position.'

Given his agnosticism about this issue, what does Rees think of the attempts which people make to calculate the probability of life on other planets? Presumably he doesn't think it can be done properly?

'No, really it can't be,' he confirms. 'Any such probability equation is based on a huge number of unknown factors. The biggest uncertainty is in the biology. For example, if somebody insisted dogmatically that the chances of life emerging on a planet were one in a trillion, trillion,

trillion, it wouldn't, at the moment, be possible to argue convincingly against the claim. The uncertainty in this biological aspect is so great that one can only be open minded about the issue. Indeed, because it is such an uncertain subject, opinion is polarised about whether there is life on other planets. Sometimes this happens in science; the less evidence there is, the more polarised opinion can become.'

It has been argued that the SETI project, the search for extraterrestrial intelligence, is frivolous, a waste of money which could be better spent elsewhere. Is this Rees's opinion?

'No, it isn't,' he replies. 'It is well worth doing. I think pretty much everybody accepts that the question of whether life exists elsewhere is fascinating; there is as wide a public interest in this question as in any other question in science. We can't lay confident odds about the answer, but I'm certainly in favour of modest efforts to search for all possible sources of life.'

There is the thought that if it does turn out that the chances of life starting up were one in many trillions then we might feel quite alone in the Universe, as if life were just an afterthought. But this is not Rees's view.

'I think it would be the wrong inference,' he says. 'If life were special to Earth, then it would certainly boost our cosmic self-esteem; the Earth, though tiny, would be the most important place in our Galaxy. It would also make us very aware that the future of life, not just on this world, but also in the wider Universe, would depend on what happens here and now on our planet.'

In terms of the here and now, Rees is not optimistic that humanity will be able to manage the various threats which inevitably accompany twenty-first-century science and technology.

'It is true that the threat of a catastrophic nuclear war has diminished,' he says, 'though, of course, there could be a political realignment at some point in the future, which might bring it back again. But the major problem in the next ten or twenty years will likely stem from the fact that people will be technologically empowered in a way that they haven't been before; for example, it will become possible for very small groups of people to make use of what is likely to be widely available biotechnology, in order to cause widespread, and perhaps catastrophic, harm. Up until now, society has always been able to tolerate a tiny number of disaffected, social misfits, because of the fact that they could only cause a relatively small amount of damage; but if it becomes possible, for example, for a single person to release a dangerous pathogen into a big city, then even one person in a million with this kind of mind-set is too many. The difficulty is that it is something which society is going to have a very hard job to defend itself against. It just seems to be an intractable problem.'

There is an argument that as society becomes increasingly desacralised, and as science and technology become ubiquitous, there will be a decline in the propensity towards the kinds of extreme irrationality which might lead a person to manufacture and release a dangerous pathogen.

'I think that this is too optimistic a view,' Rees says, when I put the point to him. 'After all, there is evidence that irrational cults and dogmatic religions are flourishing. For example, one worrying portent is the emergence of cults which are inspired by technology and science fiction. They would claim that they're not anti-science, but they're certainly a long way from being rational. We can't simply assume that people with technical expertise will be rational in the sense that we understand it.'

Are there potential technological solutions then to the problem that disaffected individuals will likely have the capability to cause large-scale harm? Perhaps increased surveillance?

'I think the public will become reconciled to increased surveillance,' says Rees. 'But apart from this, one can only hope that society will evolve in such a way that ever fewer people will be inclined to misuse their expertise in dangerous ways. I'm not optimistic, though. What really concerns me is that some kind of genetically engineered virus will be released in one of the high-density, mega-cities of the developing world. It might be that we'd be able to cope with this in the West, but in the developing world it would likely be a disaster.'

If Big-Brother-type surveillance becomes a possibility, would this be a price worth paying to keep us safe?

'Well, as I indicated, I think there will be a tendency for people to feel that their security justifies high levels of surveillance. Indeed, I'm worried about the possibility of over-reaction. For example, if there were a single incident in the United States of the kind where one disaffected biogeneticist released some pathogen into a big city, then even if the number of fatalities were controlled, I think the effect on American society would be very serious; people would be aware that if it had happened once, then it could happen again. Given that we have already seen an exaggerated reaction in America to less serious incidents involving pathogens, it is likely there would be strong pressure from the US public to introduce high levels of surveillance, to have a backlash against science, and this kind of thing.'

One response to the threats which are posed by the development of new technologies is to suggest that science might be selectively slowed down; for example, that legislation might be enacted with the view to discouraging research in particular fields of scientific endeavour. Does Rees find this idea useful?

'Well, it is certainly possible to slow down the rate of scientific progress,' he replies. 'Every scientist who has had their research grant stopped knows that removing support and facilities hinders scientific work. But whilst this is true, I don't think that it is plausible to argue that it would be possible to slow down *harmful* science and yet continue to reap the benefits of other kinds of science. Not least, one cannot predict beforehand which discoveries are going to be beneficial and which are not. If we engage in scientific research, then we have to accept that as science advances, it brings with it increased risks and greater ethical challenges.'

How does this view fit in with the position which Rees takes in *Our Final Century*, which is that public money for pure research should not be forthcoming if the outcome of the research is likely to be misused?

'Well, this is just the other side of the coin of the generally accepted practice that when it comes to funding we should favour those parts of science which are seen to have particular strategic value,' he explains. 'This is the reason that molecular biology, for example, tends to attract more funding than black hole research; they are both of scientific interest, but there is more likely to be some kind of payoff from molecular biology than from black hole research. All I am saying is that it would not be inconsistent to withdraw support from a particular line of research if we reasonably expected that it would lead to social problems. But the effectiveness of any such adjustments will be minimal; normally you can't predict outcomes and, in any case, you can't stop people thinking about whatever they want to think about. This point is of especial significance when one considers that science is now a thoroughly global activity, often driven by commercial pressures; you can't stop a particular line of research just by stopping it in one country.'

'I'm not optimistic then that we can slow down *specific* lines of research,' Rees says. 'Particularly, I think that people like Bill Joy are very naïve if they think we can avoid our problems by means of fine-grade relinquishment; by slowing down some parts of science and not other parts. It just isn't clear how we'd get agreement internationally, and even if we got agreement, enforcement would be very difficult.'

The risks associated with biotechnology are not of the kind which threaten the existence of *Homo sapiens*. Although it is possible that a pathogen released within a city of the developing world would kill millions of people and cause widespread disruption, it is extremely unlikely that it would kill everybody. However, in *Our Final Century*, Rees identifies a different kind of risk associated with some of the most cutting-edge research in particle physics. In particular, there are certain particle accelerator collision experiments which seem to involve a minute risk of destroying the whole Earth (and perhaps space as well). Although these risks are vanishingly small, Rees argues that the wider public should be consulted in order to determine whether it is considered that the risk is below an acceptable threshold. What's the thought here, I ask him?

'Well, first let me say that I don't myself worry about any of these kinds of experiments; the risks really are very small,' he replies. 'But I do think that they raise, in a stark and extreme way, issues of principle: namely, what should we do when we're planning an experiment which has a very small chance of having utterly catastrophic consequences? If the result of an experiment going wrong is that everything is destroyed, then the risk that most people would be prepared to accept would presumably be very low, in the order of many billions to one against it happening. But how could we convincingly demonstrate that the risk of such an experiment is more or less this unlikely? When you're probing the frontiers of science, when you might be encountering new physical phenomena, it is never possible to be entirely confident about the outcome of an experiment. But the wider public, with good reason, will want this kind of confidence, so it is necessary, at the very least, to keep them informed. This comes up in a more realistic way when we're talking about biotechnology; what's the right way of handling the small risk that the public will be exposed to a dangerous pathogen? So it's the normal issue of ensuring safety, but it becomes starker when we're talking about something which carries an extremely small risk of absolutely catastrophic consequences.'

Isn't there a worry about involving the general public in such debates; namely, that people will often adopt a view which is based on an emotional, rather than rational, engagement with an issue, one which is frequently led by unbalanced reporting in the mass media? The debate about GM foods in the UK, for example, is generally considered to be of a pretty poor standard.

'Well, I think that the GM debate has been strident and media led; and despite the government's best efforts, it is true that there has been very little informed discussion of the issues,' Rees answers. 'It's clear that there is a widespread suspicion of GM foods, largely, I suspect, because of the involvement of commercial interests in their production. I think the lesson to be learnt here is that although scientists ought to be willing to engage in collaboration with industry and the commercial world, there is the risk that it will be at the cost of the great social benefit of having a cadre of experts who are trusted and are seen as being impartial; and that is a real problem.'

But isn't there a more general issue here which is about a lack of scientific literacy on the part of the general public? For example, surveys in both the UK and the United States have shown that very few people know that CO_2 is the major greenhouse gas. Given this level of ignorance, it isn't clear what kind of role the public can or should play in overseeing the practice of science.

'I agree it is a problem,' says Rees. 'Indeed, part of the reason that the GM debate went badly was to do with ignorance; some people, for example, didn't realise that ordinary crops had genes in them at all.

But there have been other debates which have gone rather well in the UK: the debate over stem cells, for example. Here scientists did manage to explain the various issues properly, which meant that politicians were able to have an informed debate. And it was the same with the debate which centred on the work of Mary Warnock's Human Fertilisation and Embryology Committee.'

But in both these cases didn't the general public take something of a back seat?

'Well, I'm not sure about that, but certainly the debates didn't descend into the kind of tabloid-led sloganeering which characterised the discussion of GM foods. It *is* true that the public aren't very informed about science; however, there is sometimes the assumption that if people knew more about science, then they'd be happier with it. But there is evidence against this view. Informed people can be just as concerned about some applications of science as uninformed people. Take the debate over nuclear energy, for example. Here there is as much of a division amongst experts as amongst non-experts; explaining the details of the issue to people doesn't necessarily result in consensus. Scientists are just kidding themselves if they think that by informing the public they're going to get an easy ride. It might well work the other way around. Nevertheless, the important thing *is* to inform the public, and scientists have to accept that this means that they have the task of explaining what they're doing and responding to the challenges which might result.

'Also, it is possible to make a distinction here between the practice of science and the use of its discoveries. Particularly, it is likely that there will always be a variety of applications of any particular discovery. The choice about what to do with a discovery should be made by the wider community, not just by scientists; consequently, it is necessary for scientists to engage in a dialogue with people in order to ensure that decisions are informed. Of course, there are also going to be ethical issues involved with these sorts of choices; and it is likely these will reflect back, and perhaps constrain, the kinds of science which it is possible to pursue.'

Is one of Rees's motivations for writing popular works on science then that he thinks that scientists have a duty to keep the public informed about their work?

'That's part of it,' he replies. 'I certainly think that all scientists should do their best to make their ideas accessible, and that's not all that difficult. It's the technical details which make science hard to understand if you're not an expert, but the detail is only of interest to the specialists. The key things for the non-experts are scientific facts and concepts, and these can be put over in fairly simple language.

'Of course, it is true that there are a few areas of science which are hard to explain. Quantum mechanics, for example, is highly counter-intuitive, and it is difficult to familiarise people with its subject matter. But I don't think this matters too much. All we need to do is to make people aware that there is this theory which is counter-intuitive, and then to tell them two things: first, that it should be taken very seriously, because much of modern technology is based on it; and second, that they shouldn't be surprised that we require a theory which is counter-intuitive, because it pertains to scales which are far smaller than any-thing we encountered on the African savannah, and, therefore, our brains just haven't evolved to cope with it. If we can do this kind of thing for quantum mechanics, then we can do it for the other areas of science which are hard to understand.

'The other general point to make is that my professional expertise is in space research and cosmology, and there are two particular reasons for being motivated to inform the public about these subjects. The first is that if you're working in these fields, then you're being supported by the public to do research which has no practical short-term benefit; therefore, there is an obligation to explain to people what it is all about. The second thing is that these are subjects which hold great fascination for the public; and, for my part, I would derive less satisfaction from my research if I were only to talk to a few colleagues about it. Cosmology – along with Darwinism, dinosaurs, and so on – is perceived by the public to be interesting and non-threatening; it also helps to answer some of their basic questions. So that's the reason I'm interested in popularising it.'

Rees mentions here that his area of expertise is space research and cosmology. There are some optimistic space-travel enthusiasts who think that we might be able to escape the problems of the Earth by colonising near-space. This sounds like the stuff of science fiction, but does Rees think that it might be a plausible scenario for the twenty-first century?

'Well, I should say first of all that I don't think that it solves our problems because most of us care about the Earth, and nowhere beyond the Earth is going to be as clement an environment as even the Antarctic is here,' he replies. 'But the idea that 100 years from now there might be groups of pioneers leading an uncomfortable life beyond the Earth is not unlikely. I think it is not as far in the realm of science fiction as grey goo or superhuman robots, for example. It is something which might well happen.'

Is this possibility then a useful insurance policy against the chance that humanity will destroy itself?

'I supposes so, although actually I don't think there is much chance of the entire human species being wiped out,' he replies. 'But certainly once there are independent communities of people living away from the Earth, then it spreads the risks against the worst kind of terrestrial catastrophe. The other point to make here is that it is likely that such communities would diversify very quickly. A major difference between our near future and the historic past is that human beings, through the mechanism of things like genetic modification and targeted drugs, are going to change very quickly over the next 100 years, far faster than ever before. This means that although our perception of the possible future extends to billions of years, the horizon of predictability has contracted; we're changing so quickly that it is hard to look beyond the end of this century. We cannot even be sure what the dominant form of intelligence on Earth will be a century from now. Perhaps machines will have outpaced our intellects by then. And there could be hundreds of people in lunar bases, just as there now are at the South Pole; some pioneers could already be living on Mars, or else on small artificial habitats cruising the Solar System. Space might be pervaded by robots and intelligent "fabricators", using raw material mined from asteroids to construct structures of ever-expanding scale. I am not specially advocating these developments, but they nonetheless seem plausible, both technically and sociologically.

'Darwin himself noted that "not one living species will transmit its unaltered likeness to a distant futurity". Our own species may change and diversify faster than any predecessor, via intelligently controlled modifications, not by natural selection alone. I'd hope that genetic engineering can be stringently controlled here on Earth, though I'm pessimistic about whether such controls could be effective. But if extraterrestrial pioneers wish to mould their biology to an alien environment, then good luck to them, and to any posthuman species who eventually emerge.'

Further Reading

If you wish to explore the topics dealt with in these interviews in a bit more depth, then these books and resources are a good starting point.

Darwinism and Genes

Darwin, C. 1996 [1859], *The Origin of Species*, Oxford: Oxford University Press.
Dawkins, R. 1976, *The Selfish Gene*, Oxford: Oxford University Press.
Jones, S. 2000 [1993], *The Language of the Genes*, London: Flamingo.

Evolutionary Psychology and the Blank Slate

Cosmides, L. & Tooby, J. 2004, *What is Evolutionary Psychology*, New Haven, CT: Yale University Press.
Pinker, S. 1999, *How the Mind Works*, Harmondsworth: Penguin.
Pinker, S. 2003, *The Blank Slate*, Harmondsworth: Penguin.

The Human Brain and Consciousness

Block, N., Flanagan, O. & Guzeldere, G. (Eds) 1996, *The Nature of Consciousness*, Cambridge, MA: MIT Press.

Chalmers, D. 1998, *The Conscious Mind*, New York: Oxford University Press.
Greenfield, S. 2002, *The Private Life of the Brain*, Harmondsworth: Penguin.

Science and the Human Animal

Malik, K. 2001, *Man, Beast and Zombie*, London: Phoenix.
Ridley, M. 2004, *Nature via Nurture*, London: Perennial.
Wilson, E. 1988, *On Human Nature*, Cambridge, MA: Harvard University
     Press.

Cybernetics and a Post-Human Future

Margolis, J. 2001, *A Brief History of Tomorrow*, London: Bloomsbury.
Moravec, H. 2000, *Robot*, New York: Oxford University Press.
Warwick, K. 1998, *In the Mind of the Machine*, London: Arrow.

Psychiatry and Schizophrenia

Frith, C. & Johnstone, E. 2003, *Schizophrenia: A very short introduction*, Oxford:
     Oxford University Press.
Goldberg, D. & Murray, R. (Eds) 2002, *The Maudsley Handbook of Practical
     Psychiatry*, Oxford: Oxford University Press.
Shorter, E. 1998, *A History of Psychiatry*, Chichester: John Wiley.

Microbiology, Viruses and Their Threats

Crawford, D. 2000, *The Invisible Enemy: A Natural History of Viruses*, Oxford:
     Oxford University Press.
Ewald, P. 1997, *Evolution of Infectious Diseases*, New York: Oxford University
     Press.
Kiple, K. (Ed.) 1999, *Plague, Pox and Pestilence*, London: Weidenfeld & Nicolson.

On Cancer Research

Greaves, M. 2001, *Cancer: The Evolutionary Legacy*, Oxford: Oxford University Press.

Vasella, D. 2004, *The Magic Cancer Bullet*, London: HarperCollins.

Weinberg, R. 1998, *One Renegade Cell*, London: Weidenfeld & Nicolson.

Animal Experimentation, Ethics and Medical Research

Harris, J. 1990, *The Value of Life*, London: Routledge.

Harris, J. 2001, *Bioethics*, Oxford: Oxford University Press.

The Medical Research Council Ethics Series: http://www.mrc.ac.uk/index/publications/publications ethics_and_best_practice/publications-ethics_series.htm

Science Under Threat

Bricmont, J. & Sokal, A. 1999, *Intellectual Impostures*, London: Profile Books.

Forrest, B. & Gross, P. 2004, *Creationism's Trojan Horse*, New York: Oxford University Press.

Gross, P. & Levitt, N. 1994, *Higher Superstition: The Academic Left and Its Quarrels with Science*, Baltimore, MD: Johns Hopkins University Press.

Biodiversity

Gaston, K. & Spicer, J. 2003, *Biodiversity*, Oxford: Blackwell.

Wilson, E. 2001, *The Diversity of Life*, Harmondsworth: Penguin.

Wilson, E. 2002, *The Future of Life*, New York: Vintage.

Science: Its Dangers and Its Public

Einsiedel, E. *Public Understanding of Science* (quarterly journal), London: Sage.

Rees, M. 2003, *Our Final Century*, London: William Heinemann.

Worcester, R. 2001, *What Scientists and the Public Can Learn From Each Other*: http://www.mori.com/pubinfo/rmw/cambridge.pdf

Index